Is Theistic Evolution Viable?

Is Theistic Evolution Viable?

A Reflection on Evolution and Christian Theism

Brendan Sweetman

CASCADE *Books* • Eugene, Oregon

IS THEISTIC EVOLUTION VIABLE?
A Reflection on Evolution and Christian Theism

Cascade Books
An Imprint of Wipf and Stock Publishers
199 W. 8th Ave., Suite 3
Eugene, OR 97401

www.wipfandstock.com

PAPERBACK ISBN: 979-8-3852-6186-4
HARDCOVER ISBN: 979-8-3852-6187-1
EBOOK ISBN: 979-8-3852-6188-8

Cataloguing-in-Publication data:

Names: Sweetman, Brendan [author].

Title: Is theistic evolution viable? : a reflection on evolution and Christian theism / by Brendan Sweetman

Description: Eugene, OR: Cascade Books, 2026 | Includes bibliographical references and index.

Identifiers: ISBN 979-8-3852-6186-4 (paperback) | ISBN 979-8-3852-6187-1 (hardcover) | ISBN 979-8-3852-6188-8 (ebook)

Subjects: LCSH: Evolution (Biology)—Religious aspects—Christianity. | Religion and science. | Bible and evolution. | Intelligent design (Teleology). | Creation. | Chance. | Determinism. | Divine Action.

Classification: BS651 S944 2026 (paperback) | BS651 (ebook)

VERSION NUMBER 05/26/26

For Doug Geivett

Contents

Preface

The relationship between religion—especially Christian theism—and the vexed theory of evolution continues to provoke vigorous, often contentious, debate, giving rise to profound concerns and challenges, with exchanges frequently marked by controversy, confusion, and misunderstanding. Fundamental questions that continue to divide opinion include whether evolution can be reconciled with biblical readings and teachings; whether God's design and purpose can be reconciled with the apparent randomness of evolutionary processes; whether evolution implies that humans differ not in kind but only in degree from other species; and whether the theory ultimately threatens the objectivity of morality and undermines humanity's quest for purpose and meaning.

In this book, we take up these fascinating but complex issues with a central aim: to explore and defend the position known as theistic evolution. More specifically, we will argue that evolution and Christian theism are not merely compatible, but can be reconciled in a way that is both viable and intellectually compelling. Surprisingly little scholarship has been devoted to developing a detailed and plausible account of this position. Many existing works attempt to illustrate the compatibility of evolution and theism, yet they often fall short, especially in grappling with the tension between chance and purpose raised by evolutionary theory—one of the main themes of this book. Many thinkers approach the topic from either a creationist and intelligent design perspective, which largely rejects theistic evolution, or from a more mainstream (evolution-positive) outlook, where more energy is expended critiquing the former views, and not enough developing a plausible alternative. As far as I know, this is the first book to address directly the question of the viability of theistic evolution.

This question of viability, I contend, is the *essential question* with regard to the relationship between religion and evolution. Indeed, for most religious believers it is more significant than whether or not evolution is true. In what follows, we attempt to illustrate that not only is theistic evolution a plausible view, more plausible of course than atheistic naturalism (which is a non-starter), but that our version is more reasonable than several influential approaches in contemporary theology. From the outset, we assume the existence of God and the central doctrines and themes of Christian theism, along with the core claims of evolutionary theory (with some important qualifications, particularly regarding the role of chance). Our concern is not to investigate the truth of Christianity or evolution, but to examine whether they can be reconciled in a way that is viable. We don't aim for a strict demonstration of our position, too high a standard for the complicated and challenging area of worldviews. Yet, mere compatibility between evolution and Christian theism is insufficient; what matters, and what we seek to establish, is viability—a goal more modest than proof but one that is adequate when faced with ultimate questions of life and meaning.

For most religious believers, some version of the following argument (speaking loosely from a philosophical point of view) must be regarded as correct:

1. God exists and created and sustains the universe and all life for a particular purpose (this means that there are goals that God wishes to realize in the unfolding of the universe and life);
2. Evolution is the process by which living species come into being and change;
3. Therefore, evolution and God's purposes must be reconcilable, and so *some version of theistic evolution must be true.*

The vast majority of religious people are theistic evolutionists, so it is important to be clear about what they believe (even if they seldom reflect on the specifics of their views in a formal way). But most would hold: (1) that God created the world with a purpose in mind; (2) that God directs this purpose to a quite significant extent; and (3) that there is little or no chance in nature (though there is free will). Ordinary religious believers who would classify themselves as theistic evolutionists may differ with regard to their understanding of evolutionary processes, but the main point is that they generally deny that the process is *unguided.* Yet, the idea that evolution is chance-driven is fairly widespread in not only modern evolutionary biology but even among many contemporary theologians! Many in these circles accept the mainstream scientific view that evolution is largely governed by chance, and so in seeking to reconcile this outlook with theism, they often significantly modify—or even abandon—key Christian doctrines and themes. Moreover, although many contemporary thinkers will affirm that evolution is directed in some sense, they are often quite vague about how this is to be understood in practice. As we shall see, they often confuse or conflate real chance in the universe with apparent chance, chance with unpredictability, and never address clearly (and so come to terms with) the question of *how chance and direction/purpose can fit together and coexist*. We will explore all of these themes in what follows, as we develop our own specific version of theistic evolution.

Although our approach to this fascinating topic and cluster of questions is philosophical, I have tried hard to make the arguments and ideas accessible for a general, educated audience. In illustrating the viability of theistic evolution, we also wish to offer a plausible path forward for those many readers especially interested in considering how to reconcile Christian theism with evolution without being forced to seriously modify or distort the former or reject the latter, for the topic is not only of philosophical, theological, and scientific interest, but also carries important educational and public policy implications, not only for our schools but also for society and culture at large. Accordingly, the book is intended to appeal not only to philosophers and religious scholars but also to non-philosophers, including pastors, seminary and religious program teachers and their students, educators at all levels, and the many educated readers for whom the topic continues to hold great interest.

In our first chapter, we introduce various definitions of theistic evolution and distinguish them from creationism, secularism, and intelligent design theory, as well as providing an overview of the key concepts of Christian theism (relating to God's existence and nature, biblical revelation, and central Christian doctrines) and of evolution (such as microevolution, natural selection, chance, and macroevolution). The latter part of the chapter explains why evolution is thought to be a problem for Christian theism: as a challenge to biblical revelation; as a process governed by chance, and so a threat to the notion of design; as exacerbating the problem of evil and suffering; as undermining the objectivity of morality; and as questioning differences in kind (as opposed to differences in degree) between *Homo sapiens* and other species.

Far from being seen as a solution to the supposed conflict between religion and evolution, many thinkers reject theistic evolution altogether, holding that it is not a viable alternative to more traditional views that deny evolution or to intelligent design theory. In chapter 2, we consider three main lines of criticism aimed at theistic evolution: (1) The biblical critique: the argument that evolution contradicts the book of Genesis and even undermines biblical revelation as a whole; (2) The critique of intelligent design (ID) theorists that since evolution operates randomly it cannot be reconciled with God's design or purpose; moreover, on both logical and scientific grounds, ID is a better explanation for the complexity we find in biological systems; (3) The rejection of methodological naturalism: the argument that the commitment to methodological naturalism in scientific research is really metaphysical naturalism in disguise, and prejudices the debate toward preordained outcomes, unnecessarily and arbitrarily eliminating design. We discuss the views of Wayne Grudem, Stanley Jaki, Stephen Meyer, Francis Collins, Denis Lamoureux, Stephen Dilley, and some of the classical thinkers (such as St. Augustine) along the way.

Chapter 3 introduces and provides a detailed analysis of the concept of chance, one of the most fundamental concepts surrounding this topic. The supposed presence of chance in the process of evolution is one of the main reasons the theory has generated much controversy and rejection. We survey five meanings of chance, and then show the role that chance is thought to play in evolution and how it is accepted uncritically as a governing assumption rather than as a defensible position with regard to the metaphysical question of whether there is real chance in nature. The chapter also considers the possible ways in which God might act to introduce design and purpose into the world, with or without chance being present.

Chapter 4 turns to the argument that we live in a *deterministic* universe, the view that every present state of the universe is caused directly (or is completely determined) by prior states together with the laws of physics (and so there is no real chance in nature). We attempt to show that determinism is the most reasonable understanding of how reality operates, and is also the one most true to scientific practice. After considering a number of objections to this position, including from quantum mechanics, we turn to the implications of determinism for science, for biology and for the process of evolution, as well as elaborating further on the crucial distinction between chance and randomness, especially as these concepts relate to the process of evolution.

A theory of theistic evolution must be able to deal not only with the conflict between chance and purpose, but also to provide a plausible account of how divine action actually works in order to achieve God's purposes. In Chapter 5, we defend the position that God directly intervenes in nature from time to time by breaking or suspending scientific laws. We consider alternative perspectives from contemporary philosophy of religion and theology (that are sometimes described as "hands-off theology") based around the claims that, while God is active and immanent in creation, nevertheless God acts in an indirect way, rather than a direct way, and, moreover, without breaking or suspending scientific law. The second half of the chapter examines two influential contemporary theories of theistic evolution in the work of Arthur Peacocke and John Haught.

Chapter 6 continues our examination of alternative contemporary theological and philosophical views to the one defended in this book. We consider several influential authors who appeal to the theory of quantum mechanics as offering a possible way to develop an approach that shows how God may work within the randomness present at the quantum level to influence the course of nature overall. We examine the views of Robert Russell, Nancey Murphy, Francis Collins, and Karl Giberson, and evaluate whether they are viable alternatives when compared with our view.

Our last chapter brings together the separate threads of our previous arguments in an attempt to show that our position is a plausible account of how God guides evolution while also affirming the main features of the theory (except for the claim that it operates by chance). We discuss such key topics as the problem of evil and suffering in nature, Christian doctrines and practices (such as prayer), higher-level consciousness, reason and free will (differences in kind), and illustrate also how theistic evolution overall provides a more reasonable, plausible explanation for the objectivity of morality and better addresses concerns over meaning and purpose.

I owe a debt of gratitude to many who helped me while working on this book. I wish to acknowledge those numerous friends and colleagues with whom I have discussed the topics of the book over the years, either individually or at conferences. I am especially grateful to Doug Geivett, Curtis Hancock, and Brendan Sweetman Jr. My very special thanks to my friend and colleague, Professor Bill Stancil, who sadly passed away before the book's publication. I am indebted to Bill for providing incisive comments on the whole manuscript, and for his deep knowledge of all matters theological. I also wish to acknowledge Rockhurst University for sabbatical leave to work on a portion of the book. I am thankful also to Michael Thomson and Matthew Wimer at Wipf and Stock, and especially to Robin Parry for his many helpful comments on the manuscript. My greatest gratitude goes to my family for their always unfailing support and encouragement, without which this book would not have been possible!

Brendan Sweetman,
Kansas City,
October 2025

I

Theistic Evolution: An Introduction

Many people are convinced that the scientific theory of evolution is incompatible with religious belief, and many more seem to be quite sympathetic toward this view! The reasons inspiring this perspective include the convictions that the details of evolution seem to *undermine the story of creation* in the Christian Bible, and the claim that since the evolutionary process operates largely by *chance*, it appears to be incompatible with the belief that *God designed creation for a special purpose*. Moreover, various implications of the theory *seem* to undermine other crucial religious themes, such as that human beings differ in kind and not just in degree from other species, that God takes providential care of his creatures, that morality is objective and founded upon God's nature, and so forth. Such issues, and other related ideas, raise concerns for people on all sides of the debate, so this attitude with regard to evolution and religion is not simply an eccentric reaction that can be easily dismissed—it is held by many theologians and scholars of religion, by philosophers, scientists, various intellectuals, and by scores of ordinary religious believers and indeed ordinary atheists. From a geographical point of view, it is a perspective mostly associated with the United States, but nevertheless, it has supporters in religious, intellectual, and cultural circles in many parts of the world. Furthermore, quite a large number of intellectuals are generally uneasy about the relationship between evolution and religion, even if they believe a reconciliation is possible. A smaller number of religious thinkers enthusiastically embrace the theory of evolution as a better expression of God's creative intentions than more traditional theological accounts; many who are already atheists and secularists also welcome the theory as a way of further advancing *their* worldview!

This book is an exploration of the view usually described as *theistic evolution*. But we are not merely concerned with what theistic evolution is, and who holds it, but with a more specific question: *Is theistic evolution viable?* As noted in the preface, the question of its viability is, I contend, *the essential question* in the debate about evolution and religion, and one of the most important questions surrounding topics in science and religion as we move further into the twenty-first century. The concept of viability is a vital and indeed novel concept to introduce into the debate. Viability generally refers to whether a process or event or entity is capable of working successfully, whether a plan can function and do the task it is supposed to do, whether a proposal is reasonable, can be developed in a way that will lead to practical success, and so forth. When we consider whether a financial rescue plan for a company is viable, for instance, we are asking whether it is a plausible, realistic plan likely to save the company. We are not inquiring into whether its success is a mere possibility, but wish to know if the plan is *likely to work*, whether there is a good degree of probability for its success. Or, turning to biology, if we say that a germinating seed is viable, we are saying that in the right conditions it will very likely survive and grow, not just that there is a logical possibility that it will do so. So when we ask if theistic evolution is viable, we are asking if it can work as a successful theory to reconcile religious belief with evolution in a practical way that solves many of the questions and issues that initially prompted the worry that evolution and religion are incompatible. If it is viable, theistic evolution may be regarded then as a successful approach whenever practical living comes into contact with questions that arise from the intersection of our religious beliefs and questions that arise in evolutionary theory.

The question of viability is crucial even if many religious believers don't give much thought to evolution. Why? Because the rationality of their worldview depends upon the answer! They may *assume* its rationality in their everyday religious practices, perspectives, and mindsets toward various questions of life and responses to various experiences, including their reaction to scientific developments. This would be similar to accepting that morality has an objective theological foundation even if one never considers that foundation, or that God has a reason for allowing evil in the world even though one never contemplates any of the theories available, even when one experiences evil in one's own life. Indeed, the acceptance of the congruence between evolution and religious belief might more generally be analogous to those theists who believe in God, but do not give much thought, if any, to the rational case for God's existence. Insofar as the question might arise in their experience, they accept that there is a rational case, but they themselves are not familiar with it, could not do it justice, and seldom appeal to it in the day-to-day expression of their religious philosophy of life. Since the vast majority of people are likely to be theistic evolutionists in some form, its viability is clearly a fundamental matter, even if it is assumed by many.

Compatibility vs. Viability

So it is important to emphasize right at the beginning that we are not focusing only on whether religious belief and evolution are *compatible*. Many hold that they are, of course, but this is a weaker conclusion than the claim that theistic evolution is *viable*. It might be that we could lay out a case for how they might be compatible, but the case might not be very plausible, might require a too radical rethinking of Christian theism, or seem inconsistent with scientific evidence, at least without making (implausible) qualifications or without adding premises to the argument that are unlikely to be true. Often this is the way the topic is approached in the literature, where a general or superficial case for the compatibility of evolution and religion is proposed, but little or no specifics, or inadequate specifics, are given, so that it is not clear if the case is really plausible, or to use our language, if the position being advanced is viable. It requires more work to show that theistic evolution is viable than to show that it is logically possible. I shall argue that it is viable, and try to show why, as I develop my own view. That shall be our task in this book.

Why do we insist that the question of viability must be regarded as *the essential question* in the often fractious debate between religion and evolution? To answer, it is helpful to consider some of the various possibilities that we can delineate with regard to our topic. We could try to solve our worries about evolution and religion in several ways. One would be to reject the theory of evolution (which many traditional theists and creationists do) and so (at least on the surface) there is nothing, in fact, to worry about (*except* the not insignificant fact that one would be denying a theory that commands a large scientific consensus, a position that is intellectually costly and frequently invites scorn and ridicule). But on that approach, there is no pressing need to show that religion and evolution can be reconciled, let alone in a plausible way. Another possibility open to a theist is to accept evolution, but to develop and defend a different account of it, perhaps to deny claims that are normally advanced as part of the theory, such as to deny that the process operates with a large element of chance, or that natural selection is the mechanism of change, or even common ancestry, and to suggest that *this version* of evolution might be shown to be compatible with traditional theism. A third way is to modify, sometimes radically, what we normally mean by theism, to the extent that we can then make it compatible with evolution. For example, we might accept that God does not know the future direction of his creation, or that the biblical story of creation is not accurate, or that God never in fact intervenes in his creation, and so forth, and in this way theism and evolution can be made compatible. We will consider all of these views as we go along, and indeed our view will be a version of the second approach, *particularly with regard to the question of chance.*

It is quite common in contemporary discussions to see the question of compatibility confused with the question of viability, or at least to see the two notions conflated with each other. This is evident in much recent work, including that of religious thinkers like Arthur Peacocke, Robert Russell, Nancey Murphy, and John Haught, as well as religious scientists, such as Francis Collins and Francisco Ayala (among others), all of whose work we will discuss later. The scientists are usually less precise than the theologians, but we will see that a lot of work in contemporary theology is quite speculative and does not come to terms with the question of viability. Some thinkers talk as if compatibility is the same thing as viability, suggesting that if it is logically possible for evolution and religion to be compatible, that this is similar to saying that theistic evolution is plausible or viable, or at least they run the notions together when they should be keeping them distinct.[1]

By way of analogy, let us consider the ancient ritual of rainmaking, often practiced by performing rain dances (or other activities)—where various tribes engaged in the practice of performing a dance that was supposed, in some forms, to induce the Almighty (or spirits or ancestors) to cause rain to fall. Is the belief in the power of rain dances compatible

1. As an example, see Murray and Churchill, "Mere Theistic Evolution."

with modern science? Many will answer this question in the negative, and argue that rain dances are not compatible with physics in general, and more specifically with what we now know in the area of meteorology, citing familiar objections. These would include the claims that the dances frequently do not lead to rain, that any rain correlated with them is pure coincidence, that our knowledge of how rain occurs, even if incomplete, shows clearly that it has nothing to do with dancing or any other ritual, and so forth! In response, supporters of rain-dance theory might claim that the dances *are* compatible with modern science and respond to common objections by adding various premises to their theories, premises that seem logically *possible* (but that are quite *implausible*). For example, they might claim that there are reasons why it does not always rain when the dance is performed, including (a) the dance was performed incorrectly or inadequately; (b) dancers are often in a state of sin when they perform; (c) God only occasionally responds to rain dances, because he often has other reasons, inscrutable to us, for withholding a positive response to the dance and/or for denying rain at various times; (d) and so forth. Similarly, when it rains without the dance being performed we can rationalize that as well by adding similar qualifications to the argument, such as: God occasionally provides an unexpected gift to lift our morale, or as a reward for good behavior, not just in this case but in many areas of life; God sometimes furnishes rain for reasons we cannot discern now but often become obvious later, like forestalling a drought, and so forth. Repeated failures that lead to drought can be similarly rationalized.

The lesson of this example, of course, is that if we accept such premises and qualifications, it would be true that the practice of rain dancing could be made compatible with modern science. However, few of us, I think, would deny that this response is not a *viable* option because the additional premises face serious objections: they seem *ad hoc*; they are not plausible in themselves; they are inconsistent with the current state of the science of meteorology. This example shows that logical possibility or compatibility, while important, is not enough; we must seek to take that extra step and consider the matter from the point of view of viability. So this is our task: to consider the topic in a careful and informed way, so that we can make our best judgment about the question of viability, given that the topic will not likely be subject to any type of clear demonstrative proof. In order to get further into our topic, we now need to say more about what we mean by the key concepts in our conversation, especially relating to (Christian) theism, evolution, and theistic evolution. We need to have a straightforward understanding of the terms we are using so that we can be clear about what we are arguing.

Christian Theism: Key Concepts

Like most books on this and related topics in the Western tradition, we are primarily interested in Christian theism. This is for a number of reasons: It is the main religion in the Western part of the world; it is that religion that historically has clashed the most with the theory of evolution; it is also that worldview that has tried hardest to accommodate scientific advancements and reconcile them with well-established theological themes. So the topic of this book, at least in a theoretical sense, is one of vital importance for the dominant worldview in Western civilization, whether or not we recognize this. However, we should note as well that many of our points would apply, *mutatis mutandis*, to many of the world's great religions, that they too must be exercised by the question of the viability of theistic evolution. Indeed, it is just as essential a question for those religions as I believe it is for Christianity. It is probably true that a number of the world's religions are quite ambivalent about evolution, and so struggle with many of the issues raised in this book (though it is surely true also that, despite this, most religious believers around the world are theistic evolutionists, as already noted). So, although our focus here is on Christian theism, the discussion and arguments in this book will be quite relevant to many religions other than Christianity.

Christian theism is not quite the same as traditional theism, though the terms are usually used interchangeably or near interchangeably in most philosophical and theological literature. Nor is Christian theism, of course, the same thing as religion, though again in the Western context we often use the terms synonymously. While it is true that for our main argument we can use the terms interchangeably, we should highlight one difference between Christian theism and what we might call traditional Christian theism (or perhaps orthodox Christian theism) that some thinkers might insist upon, or where at least some confusion might arise. While there may in principle be several acceptable definitions of each term, one general difference is that traditional Christian theism subscribes to the traditional doctrines of Christianity (which we will introduce below) while Christian theism could refer to a version of Christianity that might deny one or more traditional doctrines, or perhaps modify them. A traditional theist would hold, for instance, that God's knowledge includes knowledge of the future, while a mere Christian theist might deny this. Or a traditional theist might believe that God intended for the human species to be at the top of the evolutionary tree, while a Christian theist might deny this. There are also degrees within each category. A distinguished Christian philosopher such as Richard Swinburne might deny that God knows the future with regard to free human actions, but in most other respects he is a traditional theist, whereas a leading contemporary Christian thinker like John Hick, who has more radical views of biblical interpretation and Christian theology, would not be classified as a traditional theist.[2] We will mainly discuss the topics of the book initially from the perspective of traditional theism, but it is important to keep before our minds that several of the thinkers whose work we will consider will deny some aspect of traditional theism in order to make it compatible with evolution (though, as noted, the important question concerns viability more than compatibility). The debate between the two perspectives on a number of important questions will be of central focus in our discussion.

What, then, more specifically, do we mean by traditional Christian theism? We need to highlight the key foundational-type beliefs that are held by traditional Christian theists. (Generally speaking, we will not need to get into the finer points of denominational differences.) The view of traditional Christian theism we will work with involves the following foundational beliefs:

1. *God exists, in the traditional classical sense* (God's nature has the attributes articulated in traditional philosophy and theology; God is omnipotent, omniscient, omnibenevolent, morally perfect, worthy of worship, eternal, an intelligent personal agent, the creator and sustainer of the universe, the designer of the universe, and so forth). Although this is the traditional Christian understanding of God, we should note that there is a very similar understanding in Islam and Judaism. We will not take any position at this point on questions raised by some of the attributes, such as whether God's knowledge of the future threatens human free will, or whether God is capable of change, though some of these issues may be relevant for our topic. As we will see later, it could be that one's position on the compatibility of God's

2. See Swinburne, *Coherence of Theism*, 162–78, where he argues for a modified view of omniscience; also, Hick, *Metaphor of God Incarnate*.

creating the universe and how evolution operates might lead one to deny (or modify) some of these traditional beliefs about the nature of God. (In general, one needs to be at least *aware* of whether one's position with regard to the attributes of God might have relevance for one's understanding of evolutionary theory.)

2. *The Bible is the revealed word of God to humanity.* We will stipulate that this is a key part of traditional Christian theism, that Christians accept this truth. This foundational belief is not meant to commit us to any particular theory of biblical interpretation. We acknowledge that there have always been differences among leading theologians and biblical scholars over how some biblical passages and themes are to be interpreted, but we accept that the basic Christian narrative is true and revealed by God through the prophets, Jesus, and his disciples. This narrative includes the claims that God created the earth and all life for a particular purpose, that God's election of Israel and his covenant with them occurred, that Jesus came to reconcile God to humanity, that the incarnation and the resurrection are real events, that Jesus's deity, life, death, and resurrection are central for the Christian life, that there is the promise of eternal life to come, and so forth. Crucially, this claim about revelation does not insist that the Genesis account of creation be read in a completely historical way because traditional theists have debated this matter since the earlier centuries of the church, from the time of the church fathers, St. Irenaeus (c. 130–c. 202) and Origen (c. 185–c. 253), along with St. Gregory of Nyssa (c. 335–c. 394), St. Augustine (354–430), and many other scholars right up to the present.[3] These thinkers, and others, did not deny or reject the historical narrative, but they often differed seriously over how to understand it, and so this debate was part of Christian theology from the earliest times. *This means that a traditional Christian theist need not subscribe to creationism*, the view that the creation story in Genesis must be read literally, thereby launching it on a collision course with evolution.

3. *Traditional Christian doctrines*: Traditional Christian theism also includes allegiance to other defining Christian teachings that are held to be extrapolated from the Bible and that, though sometimes subject to various interpretations, are nevertheless an essential part

3. For an overview of the many thinkers who contributed to the debate about biblical interpretation, and their views, see Jaki, *Genesis 1 Through the Ages*.

> of what is usually meant by Christian theism. These would include: belief in God's providence—the view that God creates, sustains, and guides the universe in accordance with his purposes, and ultimately looks after his people; belief in miracles—the belief that God not only can but often does perform miracles (especially as a response to prayer, a bedrock belief of Christian theism); belief in the special miracles reported in biblical revelation, such as the incarnation and resurrection of Jesus, the Eucharist, even, indirectly, the doctrine of the atonement, and so forth. As we noted, some of these beliefs and doctrines are open to interpretation (such as, does divine providence extend to free human actions?). Indeed, a Christian theist who accepted the doctrine of divine providence might not be committed to any particular understanding of it (though of course some Christian denominations do take official positions on some of these matters). There are also various theories aimed at capturing the central meaning of the doctrines of the incarnation or the atonement (of which there are at least four significant interpretations). So there is disagreement about some of these topics even among traditional Christian theists, with many scholars accepting that there is always room for clarification, refinement, and fresh insight on vital matters in theology.

So this is broadly what we mean by traditional Christian theism. On this account, a theist who denies that God knows in advance the outcomes of the natural process of evolutionary development would not be subscribing to this standard view of theism, but a theist who holds that God created the universe, but not necessarily in six days, would be. This is because the former denies a foundational claim of theism, but the latter does not, since many theologians have for centuries debated how the timeline recounted in the book of Genesis should be understood. In what follows, we will mostly refer to this view as Christian theism, or as theism, except in those cases where we need to distinguish it from specific modern views if they significantly diverge from it. So our central question then is: *Is Christian theism reconcilable with evolution in such a way that theistic evolution is viable?*

The Theory of Evolution: Key Concepts

It is unfortunate that evolution is increasingly coopted and distorted or misrepresented by various thinkers in support of other views that are not part of the scientific theory itself. As a result, the theory is often misunderstood. So it is important to be clear about what we mean by evolution, particularly about what the term includes and excludes. We are referring to the theory of evolution as it is understood in mainstream biology, which includes the following two major claims: that all species, which are constantly changing and developing over time, are genetically connected to each other in common ancestry, and that the process or mechanism of change and development that occurs within a species, and from one species to another, is natural selection. Elaborating further, the key concepts that define the theory of evolution are:

1. *Adaptation of the species*: Species change over time in interaction with their living and non-living environments, both in terms of traits and behavior, and some of these changes aid in survival through the process of natural selection.
2. *Struggle for existence*: There is a constant struggle for existence going on in nature that is often described as the survival of the fittest, in which those species (not necessarily the strongest or the healthiest) most suited for their particular environments survive best; many species (indeed, some suggest the number is in the billions) become extinct along the way.
3. *The process that brings about change is natural selection*: This is a natural process that allows species to change over time and aids them in surviving more successfully in their environments; it comes about from natural causes consisting of a combination of genetic, environmental, and, (as currently taught), chance occurrences in nature that have defining effects on the nature and development of life forms, organisms, and species.
4. *Microevolution*: This describes more localized changes within closely related species and varieties, such as within the classes of bats, finches, elephants, or whales. These are brought about by natural selection, mutations, genetic drift, and the species and varieties are related to each other genetically (i.e., the younger species are descendants of the older ones).

5. *Macroevolution* (sometimes called common descent, common ancestry, or descent with modification): The claim that all living things and species, alive now or that have ever lived, are genetically connected to each other, including not just human beings and lower animals, but also insects, bacteria, other simpler organisms, and even plants.
6. *Evolution is progressive*: The process of evolutionary change starts with simpler life forms and species, and gradually leads over long periods of time to the emergence of more complex life forms and species.
7. *Evolutionary development occurs over, and indeed requires, a very long period of time*: The theory holds that the earliest evidence of life is found in fossils, dated at more than three billion years old. Then, around 700 million years ago multicellular organisms appeared, followed at 520 million years ago by what is referred to as the "Cambrian explosion," the rapid appearance of many major species ancestral to several modern species (such as shellfish and corals). The first fish are dated to the Ordovician period (480 million years ago), the first amphibians to 380 million years ago, the first reptiles 40 million years after that, and the dinosaurs 80 million years after that in the Triassic period (250–200 million years ago). The first birds emerged in the Jurassic period (155 million years ago). The ancestors of human beings and the higher primates date as recently as the last three or four million years, with modern humans thought to have originated in a line of descent from these earlier species about 60,000 years ago. So for all of this development to take place, the theory of evolution requires that the earth be at least as old as the earliest life forms; revised recent estimates place its age at approximately five billion years (and the universe at fourteen billion years).[4]

Additional points concerning evolution:

A. *On Chance and Randomness*: Those familiar with the theory of evolution will note that there is one theme not listed—the role of chance (and the related notion of randomness) in the evolutionary process. I list it here as a kind of qualification or addition to the official

4. See Miller, *Finding Darwin's God*, 38–40, and ch. 2 for an excellent overview of the main themes of evolution. For good introductions, see Gould, *Ever Since Darwin*; Charlesworth and Charlesworth, *Evolution*.

theory because *the question of a role for chance (and randomness) in evolution, in science, and in the physical universe more generally, is very complicated and indeed is far from settled.* So while it is true that the theory of evolution, *as it is currently explained and taught*, is presented *as if* the process contains a large element of chance, it is far from clear that any events and happenings in the universe are the result of chance. We will come back to this issue below and in detail in later chapters, but I will argue in this book that evolution does *not* operate by chance; indeed, my view is that there is no chance operating at all in any part of nature or the universe. The theory of evolution as commonly taught takes what I regard as an unjustified position on a controversial philosophical, metaphysical topic and adopts what is really an *assumption* on what is still very much an *open question*. Another way to put this point is that the theory takes a controversial position on the topic of *teleology*, the idea that there is purpose in the universe and in nature. Evolution is presented as if it occurs without any teleological goals, as if there is no purpose or overall direction to the process. In this way, the role of chance in the theory, as we will see, is very significant indeed; the absence of the concept of teleology from evolution is also obviously of the first importance in considering the questions addressed in this book. Indeed, *this particular assumption* is the root of many of the serious challenges that the theory is thought to raise for the religious view of reality.

B. *Difference in Kind vs. Difference in Degree*: We should point out also that the theory takes no official position on the question as to whether human beings differ in kind, or only in degree, from other species and lifeforms. However, again, the theory is often presented *as if* it demonstrates that human beings differ only in degree from other species, another cause of controversy. The question is obviously of great significance since it is thought to have serious implications for the religion/science exchange. So we need to note that while it is sometimes argued that the theory has *implications* for this question, it is not part of the theory itself. We will come back to this topic at the end of this chapter.

C. *Subsidiary Themes*: There are a number of subsidiary themes that are not part of the theory, but that often arise in association with it, so they are not included in our main list. Such themes would include:

the origin of life; the origin of the universe; the underlying scientific laws of the universe; and the origin and nature of consciousness and related features of human life, such as reason and logic, free will and moral agency. We will return to several of these themes in later chapters.

What Is Meant by Theistic Evolution?

The term "theistic evolution" seems to go back a long way, though it has been used to mean different things. It is hard to be clear about its first appearance, but we know that nineteenth-century American botanist Asa Gray (1810–88) articulated a position of theistic evolution to express his own view of the relationship between religion and the newly proposed scientific theory, after he had become acquainted with Darwin's work.[5] It has also been employed by a host of contemporary writers to describe their own positions, though they do not all use the term in the same way.

There are at least three main ways the term may be understood today. The first meaning, a very general sense, is that theistic evolution describes the view that religious belief and the scientific theory of evolution are reconcilable. The conventional way of articulating this position is to argue that God directs evolution, so evolution is not only the way God brings about the various species (including *Homo sapiens*) but also one of the ways in which God fulfills his purposes. Most theists hold that God designed the evolutionary process, that it is aiming at particular ends, and that it is not governed by, or susceptible to, the vagaries of chance. This is the view I shall defend in this book. Although held by most theists, it is fair to say that this perspective is seldom worked out in any detail, especially in relation to evolution and the notion of chance (one of the main themes in this book).

A second understanding of the term is common in modern theology, in particular. This is the view that, although God creates and sustains the universe, the process of evolution is influenced by a large element of chance, so much so that even God may not know the final outcomes of the process. Indeed, it might be fair to say that this is a mainstream view, at least among an influential group of contemporary theologians and some other scholars, who are also at the same time vocal critics of any forms of creationism and intelligent design theory. Versions of this approach

5. See Bowler, *Evolution*, 222–28; also, Gray, *Darwiniana*.

are held, for instance, by Protestant theologians Nancey Murphy, Philip Hefner, and Ted Peters, by evangelical scientist Francis Collins, by Catholic scientists Fr. George Coyne, SJ, Francisco Ayala, and Kenneth Miller, and by Catholic theologian John Haught, among others. Collins, for instance, describes theistic evolution in this way: "God . . . created the universe and established natural laws that govern it. Seeking to populate this otherwise sterile universe with living creatures, God chose the elegant mechanism of evolution."[6] He also suggests that evolution operates with a large element of chance, even though the whole process is, in some way, under the control of God. We will see later that he is one of a number of thinkers who struggle to reconcile the notion of divine purpose with their claim that those species that emerge from the evolutionary process do so randomly.

We can identify a third understanding of theistic evolution, proposed by some theologians, such as Arthur Peacocke and Robert Russell, who have tried to develop somewhat of an intermediate position. This approach tries to chart a course between the views of those who maintain that evolution is directed and those who argue that it operates by chance—suggesting that there can be an element of *both* design *and* chance working in the operations of nature. Their task then would be to show how design and chance can work together and to consider whether chance governs overall or whether design comes out on top, in the sense that God remains in control and directs the final outcomes (and so God both controls and knows which species will emerge from the overall process). If there is too much chance throughout the universe and nature, God, logically, may lose control of the final outcomes. This view allows for a less strong sense of chance than the second view noted above because God's influence plays more of a direct role than chance, whereas on the second view, chance seems dominant in evolution. Both views differ from my own, which allows for *no chance* in the process.

We should also point out that, though perhaps not part of the official understanding of theistic evolution, most theistic evolutionists wish to hold to a theistic, as opposed to a deistic, view of God's working through evolution. This means that God is actively working in—or in theological language, is immanent in—his creation, unlike deism, which suggests that after God creates the universe he allows it to unfold without any further intervention on his part. Deism is antithetical to traditional Christianity.

6. Collins, *Language of God*, 200–201.

But, as we will see later, this emphasis on a God who is active in creation will raise key questions about the specific ways that God influences the course of events, as well as giving rise to a tension between the fulfillment of God's purposes and the presence of chance in nature.

My own version of theistic evolution differs from the mainstream view of contemporary theologians in my claim about chance. A key question that philosophers, theologians and scientists have always been interested in is whether there is any chance operating in nature, and in the universe more generally. This question has provoked much discussion in recent years, and in chapters 3 and 4 we will provide a full analysis of the topic. But let us introduce our general approach and answer to the question here. My own view is that there is no chance operating in the universe, that the universe follows a deterministic path up until at least the appearance of *Homo sapiens* (unless God intervenes directly in nature along the way, but if this happens, it is not accurately described as chance, as we will see). I believe one can subscribe to the main concepts of the theory of evolution, as listed above, yet deny that the process operates by chance, holding that evolutionary progression is deterministic in its outcomes. (This latter view appears to be denied by many evolutionary biologists, as well as by most versions of intelligent design theory, whose proponents often *define* evolution as operating by chance, as we will see.) In saying that biological evolution operates in a deterministic way, just as physics does, we are not so much saying that God designs the path it takes, but that it follows a deterministic path *after the process begins*. This means, as we will see, that given the initial ingredients that exist at the beginning of the process of evolution, together with the laws of science, the outcomes *must* occur (at least until the appearance of *Homo sapiens*).

The argument that evolution operates by chance is, I contend, a controversial claim, and may be regarded at least in some instances as an exaggeration, rather than as an obvious feature (based on a reading of the evidence) of the theory of evolution. Even if some insist that it must be regarded as part of the official theory, given its history, I hold that this is a mistaken reading of how nature works. It is not based on scientific fact or practice, but is a way of reading or interpreting the theory (and nature's operations in general) *which is far from settled, and which should be acknowledged as the open question it is.* When we describe any causal process in nature, we should always ask if the process came about in a deterministic way, or might it have contained an element of chance (and so not have occurred), whether the process occurs in the domain of physics or biology. Just as the answer in physics is no—there is no element of chance—so should it be in biology, I contend. The discipline of biology is not entitled to the assumption—and it is an assumption—that biological causal processes operate with a (large) element of chance.

So my view differs from the mainstream theological understanding of theistic evolution described above in that it does not allow any role for chance. It also differs from more traditional views in biology and theology in a related, important sense: I hold that design is therefore clearly detectable in nature *because of the deterministic way that nature operates*. It is not detectable only in evolution, since all nature operates by deterministic laws (not just biology). But it is quite evident in biology as well, once one considers the fact that complex species, such as *Homo sapiens*, emerged from the process. Indeed, design is at least as *evident* in biology as it is in any other part of the universe. In saying that design is detectable in nature due to the deterministic way nature operates, we can distinguish our view from intelligent design theory (as we will explain later), since the latter view claims that design is detectable in the overall structure of biological systems (a view I am sympathetic to, but do not argue for here). As our discussion unfolds, however, we shall also consider the position that the evolutionary process does operate by chance, illustrating fully what this means, examining its consequences, and comparing it with our own take on the matter, as well as discussing whether it is a plausible view. This is an important topic in the overall conversation because the view that evolution operates by chance is so widely held. We need to examine fully why this is the case, what kind of arguments are advanced to support this view, and what the consequences would be if we were to abandon it, as I argue we should.

This is the version of theistic evolution that I will defend. So let me emphasize again that for us in this book, "theistic evolution" means that the way God created the variety of living species was through the evolutionary process operating according to God's design, and with no chance occurring anywhere in the route and progression of evolution. My argument is that this version of theistic evolution affirms not only the compatibility of theism and evolution, but also that it is a *viable* way of reconciling the two positions. I will also try to show in later chapters that the other two alternative forms of theistic evolution face serious difficulties and so *are not viable*.

More specifically, then, by theistic evolution we shall mean an acceptance of the following beliefs, which define the version we are defending:

1. God exists and created the universe and all life for a purpose, with human beings at the top of the evolutionary tree by design, not chance.
2. We accept current theories on the age of the earth, which suggest that the earth is very old, but acknowledge that the current estimate of how old is somewhat speculative.
3. We accept that evolution occurred, according to the best work in current science—though we *deny* that the process contains any element of chance or randomness (that is our current view, while acknowledging that the question is at very least *an open question*). This means that God is directing not only evolution, but nature, in a deterministic way (to be explained in chapter 4). I will argue that *the view that there is no chance is not only logically and scientifically more accurate, but that it also makes the reconciliation of evolution and theism more plausible, and so more viable.*
4. Possible additional beliefs: Theistic evolution might include other theses not necessarily part of its core—such as that God occasionally intervenes in nature, that God never intervenes in nature, that God knows all that will occur in the future, that miracles are possible, that science cannot address questions regarding morality, free will, and the origin of the universe. Positions on some or all of these topics are usually part of most versions of theistic evolution, but are not essential to it. We will develop all of these points as our discussion unfolds.

Theistic evolution differs from creationism (or creation science) with regard to two key points: First, it accepts the theory of evolution (keeping in mind our discussion above about the place of chance in the process), and second, it does not insist that the Book of Genesis be read as an accurate factual account of historical events.

Theistic evolution differs also from most forms of intelligent design theory because it accepts that natural selection is the mechanism of change in the process of evolution (again subject to our point about chance); it also usually denies that God intervenes directly in nature at certain specific points in time to guide the evolutionary process along a certain path, and denies that God took a direct hand in the design of some biological systems, two key claims of most forms of intelligent design theory.

Theistic evolution also clearly differs from atheistic naturalism (a worldview often known these days by other terms, such as philosophical atheism, atheistic reductionism, materialism, scientism, secularism, Darwinism).[7] This is the view that everything that exists is physical in nature, consisting of some configuration of matter and energy, and that evolution, operating with a large element of chance, is how life and the array of species came about. According to this outlook, the universe follows "blind" laws of science since there is no God, no overall designer; the physical universe basically constitutes the whole of reality. Atheistic naturalists usually appeal to evolution to support their general worldview, which is why this outlook has become inextricably linked with the theory in recent decades. (We shall have more to say about this view later in this chapter.)

Theistic Evolution and Science

One might wonder whether theistic evolution is itself a scientific theory, or whether it is subject to scientific verification and oversight? Although theistic evolution includes the scientific theory of evolution, it is obviously not a scientific theory itself; its aim is to explain how evolution and religious belief can be reconciled with each other, not to explain a particular process that occurs in the natural world. Second, it is not subject to the usual application of the scientific method—that is, it is not a theory that can be tested in an empirical sense, or from which we could make specific predictions about the natural world. Third, if we call theistic evolution a theory or perhaps an (inference to the best) explanation, it is one that takes central aspects of Christian theism and that accepts the general claims of the theory of evolution (as noted above) as true, and shows how they can be reconciled in a theory of reality. In this way, we can describe it as an explanation that combines theism and evolution, and, as I pointed out earlier, of all the theories on this topic, this is probably the most widely held one among religious people.

However, fourth, we must say more than this—and we can see what more we need to say by asking whether or not theistic evolution is a worldview. It is clear that evolution is not a worldview, of course. (We will talk a little about the connection between the secularist worldview

7. For an excellent critical introduction to this view, see Goetz and Taliaferro, *Naturalism*.

and evolution in a moment, and how they are often confused with each other.) Since Christian theism is a worldview and evolution is not, it might initially seem as if we are comparing apples and oranges. But the best way to understand the relationship between theism, theistic evolution, and evolution (or indeed any scientific theory) is to say that (our understanding of) Christian theism is that it is a worldview, and that part of it includes: first, a strong commitment to the search for knowledge in such disciplines as psychology, economics, mathematics, literature, philosophy, theology, music, *and science*; second, a commitment to follow the evidence wherever it leads; and (therefore) third, a commitment to evolution as supported by the evidence. The Christian view, therefore, is also committed to (or at least must take account of) various theories proposed in the academic disciplines, and if any of these turn out to be controversial, one might need to take a particular stand on one or other of them (e.g., in the areas of economics, psychology, education, or gender studies). So, in this general sense, *theism* is the main position to which the religious believer is committed. Then, *within this position*, theistic evolution would be one of the subsidiary theories held. It would be a way a Christian thinks about or illustrates the coherence of God's creation and the process of evolution. *It is a way of showing that Christianity has a viable theory of the creation and development of life that is consonant with science.* So theistic evolution then is not part of the definition of theism, any more than (theistic) electricity or (theistic) space geometry is part of the definition of theism. Though we should acknowledge that the topic of the origin and development of species generally has more significance for theism than does the movement of electrical particles or the mechanisms of movement in space!

These considerations also show that it is not appropriate to refer to the view we are discussing as evolutionary theism, or evolutionary reality, or evolutionary religion, or evolutionary creation, or even as a religion of evolution. Evolution is only a part of the Christian worldview—though, admittedly (from a philosophical or metaphysical point of view) a significant part—yet it is not accurate to define theism by adding evolution as an identifying descriptor. Evolution does not have much practical or spiritual significance for most Christians. The most reasonable approach is to recognize that evolution is one (of many) theories that fit within the larger Christian worldview. Indeed, in terms of Christian living, it would be a quite small, even negligible, part of the Christian outlook, and would exercise people mostly at a theoretical level, and only a small number at that. Is it possible to reconcile other scientific theories with Christian theism, such as electrical theory, photosynthesis, string theory, or even the theory of relativity? Of course it is, but these theories do not unduly press on us, because they do not usually raise questions that place the respective theory on a collision course with theism (this is true even of relativity theory, a dramatic thesis with far-reaching consequences for such topics as the nature of time, the relationship between time and velocity, even the absolute place of things in the universe. It is a theory that could perhaps in the future give us a different perspective on things that many of us take for granted).

This brings into focus an important distinction relating to the nature of, and the reach of, science; between what I call the realm of the rational (or the reasonable) or the domain of the logical, and the realm of the scientific (understood in the modern sense of being restricted to the empirical study of nature). This distinction is important since there is an unfortunate tendency in contemporary culture to conflate science and rationality. To see the mistake here, we need to remind ourselves that many arguments can be rational without being scientific. For example, standard philosophical arguments for the existence of God are, according to many, reasonable or rational arguments; the argument for the immorality of murder is a rational argument, but not a scientific one; the argument that democracy is morally preferable to communism is a rational, but not a scientific, argument. Many serious thinkers have defended the system of democracy; their case is built on logic and reason, but not generally on the application of the scientific method, which involves gathering empirical data, performing experiments, testing hypotheses, and so forth. Similarly, with regard to the case for Christian theism, which is often presented as a kind of inference to the best explanation of all aspects of reality, taken as a whole.[8]

Sometimes we overlook the fact that the realm of the rational is broader than that of the scientific. Of course, occasionally the line may be somewhat blurred, so much so that it can prompt a discussion about the definition of science—where the boundaries of science end, and where rational argument takes over. We can be intimidated by the success of science, so that we forget that there is nothing mystical or earth-shattering about the scientific method—it is a quite tedious, prosaic process of

8. One of the foremost exponents of this approach is Swinburne who has written a series of books laying out the rational case for Christian belief; see especially: *Is There a God?*; *The Existence of God*; *The Resurrection of God Incarnate*; and *Revelation*.

study and investigation based on the application of ordinary reason in the domain of the physical. Its subject matter involves an ordinary commonsense approach based on observations, gathering of facts, occasional experiments, etc., and then logical arguments and inferences based on the evidence found. It can be a quite mundane, even monotonous, way of studying the physical realm (consider studying the long-term effects of marshland on the growth of trees), even though its results are often very impressive. And so we must keep in mind that scientific reasoning is nothing but the application of reason and logic in a specific domain, and that science is only a *subset* of the realm of the rational, not a definition of the rational.

Science and Secularism

Just as it is important in setting up our topic to distinguish between the realm of the scientific and the realm of the rational, so should we distinguish clearly between the discipline of science and the worldview of what we will refer to as atheistic naturalism. It has now become a commonplace in considering the relationship between religion and science more generally to distinguish between science and atheistic naturalism (or philosophical atheism), or perhaps more specifically between evolution and atheism. The confusion between science and atheistic naturalism has been one of the worst distractions in the modern controversy, and is associated with misleading and tendentious positions such as atheism masquerading as science, and the attempt to promote philosophical atheism by appeal to science. All of which creates an unfortunate suspicion in the minds of the general public with regard to science, among other problems of confusion, misunderstanding, and ignorance. Atheistic naturalism, as we noted above, is the view that all aspects of reality are physical in nature, and so can be explained, at least in principle (if not currently) by means of science and the scientific method. It has become a popular contemporary approach that leading atheistic thinkers have propagated over the past generation or so. The motivation behind the rise of atheistic naturalism was to give a more positive spin to the general atheistic worldview, thereby garnering, it was hoped, a more positive outlook and respectability for that view.

Traditionally, atheism could be looked upon as mainly a negative approach, in the sense that, oddly from the point of view of practical living, it was based on a denial and rejection of religious approaches to reality, meaning, and morality (and so it defined itself largely by rejecting the existence of God, religious teachings, religious leaders and churches, and large parts of religious ethics). But such a negative approach is obviously unsatisfactory from a logical point of view, since eventually one must decide what one believes in a positive sense (and not just what one rejects), and so naturalism was gradually developed to meet this need, and to try to give a more intellectually respectable face to atheism. It promoted the view that the physical realm is all there is, a thesis that would encompass human beings and their properties (such as consciousness, free will, and moral agency). And some atheistic and secularist groups began to put out manifestos drawing attention to and promoting their beliefs with regard to moral and political topics.[9] So gradually this view developed into a more obvious, though still incomplete, *secularist worldview*. Thinkers following this line realized that they would eventually need more complete theories of the human person, morality, and politics, in addition to an embrace of the latest scientific approaches to evolution and the nature of the universe.

This view is now more widely promoted by a cadre of well-known intellectuals, who have significant followings and readerships. This group would include Richard Dawkins and Alex Rosenberg, who emphasize the scientific side of things, focusing on evolution and on science to explain most aspects of reality. Other thinkers, such as Carl Sagan and Steven Weinberg, focus on the structure of the universe and the attempt to explain it fully by means of science. Some thinkers, such as Steven Pinker, Christopher Hitchens, and Sam Harris, stress politics, morality, and culture, and are less interested in science and evolution, though crucially, their full theories would require an amalgamation of the two.[10] The latter thinkers are often described as secularists, since secularists tend to focus on morality and politics, and wish to move culture in a secularist direction and away from religion. Naturalists, on the other hand, have less interest (usually) in politics, and are more concerned with cosmological issues, the nature of space and time, causation and chance, and the basis of cosmic and biological change. For thinkers in both groups, evolution in an essential part of their worldview, regarded as necessary to shore

9. For more on this, see my *Why Politics Needs Religion*, 59–85.

10. Recent works promoting this view include Dawkins, *Blind Watchmaker*; Sagan, *Cosmos*; Rosenberg, *Atheist's Guide to Reality*; Pinker, *Enlightenment Now*; Harris, *End of Faith*; Hitchens, *Portable Atheist*.

up their general atheism; for them, it is "the only game in town," as Alvin Plantinga puts it.[11] This is why they push very hard on their understanding of evolution (particularly the claim that evolution operates by chance), and are often unwilling to disentangle it from the larger religion/atheism debate, and also why they are sometimes accused of exaggerating the evidence and downplaying gaps and other problems facing the theory. This general attitude, of course, only feeds into the suspicion of their critics, whereas a more open mindset toward the deeper questions would surely be more fruitful. Their views are opposed by religious believers of all stripes and denominations, since, no matter how traditional or liberal theists might be, they would still reject any form of the naturalistic/secularist worldview, with its concomitant denigration of religion.

So the distinction between atheism, naturalism, secularism, on the one hand, and science and the scientific method, on the other hand, which had once caused much confusion (especially among the general public), is becoming more familiar, and is also recognized as an important, indeed essential, distinction for our topic. As rational thinkers, we need to respect science, and to look at the evidence for any theory in a reasonable way. Yet, this does not mean we must accept theories *unquestioningly*, nor should we be intimidated into accepting some of the more controversial scientific claims at face value, nor too easily acquiesce to contentious philosophical claims (such as about chance and purpose) as if they are settled scientific views. So, especially in the present debate, we must always be mindful not to be intimidated by peer pressure in the academy and not to be intimidated by science. The latter posture is surely detrimental to a proper examination of a whole range of fascinating but complex questions.

What Is the Problem with Evolution?

Let us turn now to the central question of why it is that some consider evolution and religion to be incompatible, and therefore by extension that theistic evolution is not viable. What arguments do thinkers proffer that suggest an incongruence between evolution and Christian theism? This question also requires us to look more closely at the claims and data of evolution itself, to see what it is about these claims and data that suggests evolution might not be reconcilable with religious belief. It is a fact

11. Plantinga, *Where the Conflict Really Lies*, 24.

that some people look at some of the claims of evolution, which they regard as true, and think that such claims are not compatible with, or are not likely to be compatible with, any form of Christian theism (or indeed with any form of religious belief). So there are several issues to keep an eye on: (1) Which parts of the theory of evolution are supposed to be incompatible with religious belief? (2) Are these claims essential aspects of the theory, or do they involve extending the theory controversially into areas that perhaps it should not be extended into? (3) Are the worries about incompatibility entirely motivated by a dispassionate appraisal of the evidence, or are they motivated by a desire to *make* the theory incompatible with religious belief? The latter might seem like it would be an unlikely occurrence; however, it is a stance that is, in fact, widespread in the contemporary debate and mirrors almost exactly the frequent questioning by critics of religion of the motives of those who are critical of the theory of evolution on scientific grounds.

There are at least five areas where evolution is thought to raise questions for traditional Christian belief:

First: The most obvious way evolution is thought to clash with theism is that it is regarded as a challenge to the reliability of the Bible. It is a challenge in two ways. The first is as a direct challenge to the biblical story in Genesis. If the biblical story is true, the argument goes, then evolution cannot be true, since evolution describes a completely different way that species came into existence and developed over time than the Genesis account. But, the argument continues, since evolution is true, the biblical story must be false, and so the Bible is not fully accurate, is not reliable, and is certainly not inerrant (i.e., without error). The Genesis story of creation need not be read in an exact literal way for this argument to work; one only needs to insist that the biblical account is supposed to be reporting historical facts, and that these facts are not compatible with what the theory of evolution claims to be true.

Another concern involves extending this argument a bit further. Some thinkers went on to suggest that if the creation story in Genesis is not true, then the Bible *as a whole* is probably not reliable (indeed, some religious thinkers worry about this as well). It is not just that this or that story is not reliable, that some historical claims in it may not be accurate, or that some events might not have taken place just as described, but the whole narrative is not trustworthy, including key doctrines in the New Testament (such as the incarnation and the resurrection). Therefore, Christian theism as a whole seems to be undermined. Those who think that Christianity must be based on the truth of the Bible, or on the reliability of the resurrection event (which would be most Christians, as well as most Christian denominations), it is claimed, must accept that if the Bible is not reliable, then their faith is undermined. Historically, this was a main worry that both supporters and critics of Christianity clashed over. We will discuss both of these concerns in our next chapter.

Second: The next worry is more abstract, complex, and controversial, but nonetheless a real concern for all that. It is the claim that a consideration of the evidence for evolution shows that the process operates largely by chance. We will discuss more precisely later (in chapters 3 and 4) what "chance" (and the related term "randomness") mean as we delve further into these topics in order to develop responses to them, and to explain our own position. For now, we wish to identify the main point of the worry that chance operating in evolution is supposed to cause for religious belief. The main concern is that if the process of evolutionary development throughout nature, and particularly the operation of natural selection, works with a large element of chance, it follows that the specific paths that evolution took—leading to the coming into being and the development of all species—*did not have to occur*. According to this interpretation of the theory, it is a matter of chance which pathways evolution traveled down, and so which species came into existence, and even which features a species developed during the course of its evolution. It follows that no species *has* to come into existence, and no species exists necessarily, including *Homo sapiens*. Moreover, most likely the biological features we ended up with, such as our bodily structure and systems (digestive, respiratory, circulatory) came about by chance; all these could have been quite different if evolution had taken a different path, and it could have, since it operates largely due to chance. This conclusion can also be extended to what are often regarded as our supra-biological features, such as our general level of intelligence, our spiritual and moral dimensions, the range and depth of our emotions, natural attitudes and mental traits, and so forth. Many of these features are often presented by more traditional religious thought as belonging to the human essence. However, given that evolution operates by chance, it would mean that this "essence" could have been different. This fact undermines any claim that humans have an essence at all, because an essence usually implies a fixed nature that defines a being, that came about by necessity (and, importantly, that cannot be significantly changed going forward). This idea is central to much of traditional and medieval philosophy, but has been strongly rejected by contemporary thought, often, though not exclusively, under the influence of evolutionary theory.

The upshot of the role that chance plays in the process of evolution, then, is very significant. It is thought to undermine the presence of any design in nature (as we noted earlier), to undermine the claim that God or an Intelligent Mind might have directed evolution or designed the process so that it would lead intentionally and progressively to the evolution of humanity as the highest species. If evolution, in fact, operates by chance, then all species came about by chance, and not design, and this, it is argued, represents a severe blow to all versions of Christian theism (indeed, for any religious view). This objection is also sometimes aimed at the doctrine of divine providence, and indeed is one reason that some traditional theists reject evolution. This topic will be the subject of extensive discussion in chapters 3 and 4.

Third: There is a further worry, which relates to the problem of evil. The problem of evil is the problem of why an all good, all knowing, and all powerful God would allow both natural and moral evil (evil done by human beings) to occur in the world. Such evil leads to much suffering and seems incompatible with God's nature, and so since moral and natural evil occur, the argument continues, then God does not exist, or probably does not exist. Although this is an ancient problem, and predates the theory of evolution by many centuries, it is claimed that the evidence from studying evolutionary processes makes this argument stronger, and so the case against theism becomes more convincing. Evolution shows that there is a vicious struggle for existence going on in nature between various species (one of the key themes of evolution), a struggle that involves a predatory and parasitical battle that leads to much suffering, hardship, and death. It is a fact of nature that species live in competition with each other, and that all species, including *Homo sapiens*, are subject to the vagaries of the evolutionary process (such as the arrival on the scene of a new disease like the coronavirus), which leads to much suffering. So the theory of evolution shows that there is a realm of suffering and hardship built into the very heart of nature; it is not just a side effect, an irritating but manageable peripheral feature, but part of the central engine of evolution. It might be going too far to say that evil is an essential part of evolution, but nevertheless, it seems to be a major driver of the process and so further evidence against the existence of a designer. So the overall argument is that the theory of evolution makes the problem of evil much worse than it was before evolution (and it was bad enough) thereby strengthening the case against the Christian God. We will address this question in detail later, and especially in chapter 7. We will also see that this worry is one motivation driving those contemporary theological accounts of theistic evolution that allow a large role for chance in God's creation.

Fourth: Some thinkers interpret the theory of evolution as an indirect argument against the objectivity of morality, a bedrock aspect of Christian theism, and indeed of most religious worldviews. Although evolution itself does not address the issue of morality directly (or related issues such as consciousness, rationality, and free will in human beings, as well as the development of human culture), nevertheless the theory is thought to have implications for the topic if developed to its logical conclusion. Although evolution is generally thought of as a theory about the development of the physical aspects of a species—the skeletal, biological, and chemical nature of organisms and so forth—the argument is that other, more intangible aspects of especially human life—such as culture, morality, and community living—must in the end admit of evolutionary explanations as well. Just as the theory claims that the physical structure of human beings evolved over many millions of years, so too did the intangibles evolve as part of, or alongside, the physical aspects. So human culture would have evolved at the same time as the human species evolved. So too would morality.

How does this argument, if true, undermine the objectivity of morality? The worry is that since the process is governed by a large element of chance, and is not aimed at any particular end or goal, has no purpose in mind, then the moral code that emerged from the process *could have been different* (in the same way the species could have been different). But this means that those current moral values that we cherish the most (for example, freedom, equality, justice, natural rights, truth) are products of an evolutionary process that was not aimed at them, not aimed at bringing about a species that would value them. The moral realm we have settled on for now occurred only by chance. It does not necessarily follow that we should not be guided by, or live by, such values (since they would still have a strong psychological grip on us), but it would follow that such values are not necessary. They are not part of God's order, not influenced by God's nature, not part of any objective moral order that the universe and nature have to arrive at or be guided by. Such values emerged almost entirely by accident, and so a different set might have come into being from the process had it gone differently, and of course, it *could* have gone differently.

Inevitably, these worries do seem to threaten the foundations of objective moral values, and perhaps also our motivations to adhere to them.

There is a second feature to the possible implications of evolution for the moral order. This had to do with free will. While this argument is not strictly a consequence of evolution as such, it often accompanies extended thinking about the theory, and some thinkers indicate that it was evolution that inspired their position on free will. Free will may be defined as the ability of human beings to make a real choice between genuine alternatives in such a way that the choice is not caused or predetermined by prior causes of a biological and/or psychological and/or environmental nature. It is thought to be a special feature of humanity, and traditionally was regarded as an argument against all forms of reductionism or materialism concerning the conscious mind. An atheistic interpretation of evolution is often accompanied by a naturalistic or a materialistic view of humanity. If one holds to a materialistic and reductionist view of reality, including human reality, then one is often in a position where it seems that one has to (sometimes reluctantly) deny the existence of free will, as just defined. There seems to be no room in a materialistic view of reality for freedom of this type, since everything is subject to cause and effect, including the human brain (which, though more complex, would operate just like any other organ, such as the heart). So when human beings make a decision, although we may believe it to be free, and may assume freedom on the part of others with whom we interact in the course of ordinary living, all actions of this type are in fact determined.

The implications of such a view for morality are obvious; morality would not be possible since it is dependent upon the belief that human beings act freely. Similarly, the concepts of punishment, responsibility, and even freedom in the democratic sense of the term would all be undermined if we accepted such a conclusion. This matter is connected with the theory of evolution because it is often held that evolution shows us that humans are purely physical beings, that even those aspects of humanity that are usually thought not to belong to the physical realm—such as consciousness, rationality, free will and moral agency, our spiritual nature, and so forth—all likely evolved by means of the same process as the physical body. Moreover, even if such features are not physical themselves in an underlying sense, they still depend upon the physical for their existence. So if we follow such a view to its logical conclusion, it would mean that these aspects of our humanity are not special, but originated in the same way that every other feature originated. This strongly suggests, critics argue, that we have no free will, since there is nothing outside the cause-and-effect physical operation of nature (this is one area where the contradiction between *this* claim and the claim that evolution operates by chance starts to arise).[12] We will return to this topic in chapter 7.

Fifth: While not directly addressing the question itself, many thinkers argue that the theory of evolution suggests that human beings differ only in degree from other species, and not in kind, thereby undermining the idea of human uniqueness. Traditional Christianity has always held that human beings are the highest intended outcome of God's plan for creation. This places us in a different category from the rest of creation, supporting the view that we differ in kind from other species. This difference in kind is not just based on *Homo sapiens* being the intended outcome of God's design plan, but also on the fact that we possess enormously significant characteristics or properties that other species lack, such as higher level consciousness, which includes understanding, rationality and logic, as well as free will and moral agency, and most important of all, a spiritual soul that is above the material order. The worry is that evolution challenges this general view of humanity because it suggests that all living things are on a continuum (shown by the evidence for common ancestry), rather than humanity being placed in a special category and all other living things in a separate grouping. This argument is accompanied by further attempts both to explain our special human traits, such as rationality and understanding, in biological, material ways as the product of evolution, along with a corresponding attempt to elevate the status of primates and other life forms in terms of their language, reasoning, and other capabilities, in an attempt to narrow the gap, as it were, between us and the apes. Evolution, it is argued, places this concern on the table in a way that it was not before the theory came along. While it does not show that humanity differs only in degree from other species, or even whether it really matters one way of other for deeper questions of origins, understanding, and meaning, it is still a challenge that seems to require a rethinking of the traditional theistic outlook (a topic we will also return to in chapter 7). But our next step must be to consider why it is that many religious believers reject theistic evolution in any form.

12. See Ruse, *Taking Darwin Seriously*, 253; also, Dawkins, *River Out of Eden*, 131–33.

2

What Is Wrong with Theistic Evolution?

Many readers will be familiar with the perspective that evolution and religion are not compatible. If you, like me, are a supporter of theistic evolution, you have probably been confronted with the argument that one must choose between one or the other. A professor once expressed the view to me at a symposium that even though many think so, evolution, once one fully understands it, cannot in fact be reconciled with religion in any form. A minister warned me in the strongest possible terms that any kind of acceptance of evolution, however well-intentioned, has the consequence of undermining the Christian gospel. One implication of the theory, an atheistic naturalist boldly claimed after our panel discussion had concluded, is that it fatally undermines the religious worldview (albeit slowly, he allowed, in cultural terms)! Many on all sides just won't buy the argument that evolution and religion can be regarded as compatible, let alone that theistic evolution might be viable. So in this chapter we will consider the critical attacks typically aimed at theistic evolution, not by secularists and atheistic naturalists, but by creationists and other theologians, and their fellow travelers. Those who reject the theory of evolution thereby reject theistic evolution and are often very sympathetic to intelligent design theory (ID). However, many of the difficulties we shall discuss in this chapter are advanced by secularists and atheistic naturalists as well. Though coming from the polar opposite perspective, these critics have one thing in common with ID theorists and creationist supporters—they believe that evolution and religion are incompatible.

Theistic critics of evolutionary theory include leading thinkers in evangelical and traditional theology as well as the foremost thinkers in intelligent design theory, among others.[1] Although often dismissed by the broader academic community, this is in fact a group of highly educated and talented intellectuals who have produced much good analysis and profound discussion of this topic. In addition, especially in recent years, and particularly with regard to ID theory, they have made an enormous contribution to our understanding of such topics as teleology and chance in biology, the definition of science, identification of and challenges to scientism and naturalism, God's action in the world, and critical examination of the evidence for evolution. It should be acknowledged that we benefited enormously from the scholarly exchanges between opposing sides in this general dispute.

So we need to consider carefully the objections of this group of thinkers to the view we are defending in this book and not glibly dismiss them as if they are not worth taking seriously. Another key point about this group that should not be overlooked is that they are in fundamental *agreement* with theistic evolutionists like me on what are regarded as the main points of contention in the debate with atheistic naturalism: that God exists, that God is the designer of all creation, and that the Bible is the revealed word of God. Most theistic evolutionists agree with these three central claims (at least in some form). Our differences with this group of scholars, then, emerge in how these claims are to be understood and fleshed out, and here there are major disagreements. Nevertheless, it is important to emphasize, lest we overlook the point, that there is more agreement on the overarching, defining claims among theists than there is disagreement. Perhaps it would be going too far to describe it as a friendly disagreement among ideological fellow travelers, since some of the disputes are rather more than that, but nevertheless, I suggest that a more irenic approach is called for as we move forward. This is not only because of the strident nature of the secularists who, to invoke an appropriate phrase, place all of their faith in a scientifically based naturalistic worldview, but because the issues, though complex, are of such interest

1. For a good overview of typical criticisms, see Moreland et al., *Theistic Evolution*; also Meyer, *Signature in the Cell*; Dembski, *No Free Lunch*; Ham et al., *Four Views on Creation, Evolution, and Intelligent Design*.

that an honest, open, mutually enriching discussion seems to be the only way to make progress. Few questions or positions should be closed off, or dismissed as not worthy of discussion, in this multifaceted debate.

I will classify the criticisms that the creationist/ID thinkers aim at theistic evolution into three broad categories, while recognizing that there may be some points that don't fit neatly into them and acknowledging that there may be other categories as well. But these three are, I believe, the best way to organize and understand the main points of contention. Just before we do that, we need to be clear about how this group of critics understands the position of theistic evolution. We recall from the previous chapter that our own understanding of theistic evolution is that God designed the universe, that evolution is the way species come about and how they change, but that chance is absent from the process. This is not the usual understanding of evolution presented by our scientific authorities and textbooks, which see it as operating *by chance*, without any *teleology* involved. Theistic evolution, in most of its contemporary forms, is usually accompanied by this understanding of (scientific) evolution (though sometimes the issue is fudged in the work of some thinkers, as we will see). The important point to stress here is that the creationists/ID thinkers focus their criticisms mostly on this latter understanding of evolution. They are especially hostile to the view that evolution operates randomly and to the implication that there is no teleology in nature. They also tend to hold that evolution *must* be understood (even defined) as a process that occurs by chance, a claim that oversteps the mark in my view, but a main reason why they are reluctant to accept the theory in any form. With that in mind, let us introduce the three main critiques of theistic evolution developed by the creationists/ID theorists and then examine each in turn.

Overview of Creationist/ID Critiques

1. **The biblical critique:** Let us refer to the first set of objections as the biblical critique. This position has been very well articulated by Wayne Grudem, among others.[2] Grudem identifies some of the main points expressed in the Book of Genesis, chapters 1–3, and argues that they are not compatible with any version of (theistic) evolution. Moreover, he argues that if we allow evolution to call Genesis into question, it seems

2. See Grudem, "Incompatibility of Theistic Evolution."

to follow that the whole of the biblical texts may be undermined. The essential claim motivating the biblical critique is that the creation story in the Book of Genesis does contain what are meant to be regarded as straightforwardly historical reports of events that actually occurred. If this is so, then the argument is that the theory of evolution is incompatible with most of the historical claims made in the early part of Genesis. So if we are to accept traditional Christian theism, we cannot acquiesce to the theory of evolution. A further claim is usually added to this argument: If we "reinterpret" the creation story in order to accommodate an evolutionary account, we run a risk of undermining other parts of the Bible (particularly if we are seduced by the hegemony of modern science over contemporary intellectual life). What is to prevent us from concluding that the resurrection is also incompatible with modern science and then reinterpreting it in a metaphorical way (as indeed some Christian thinkers have done)? Underlying these kinds of arguments is sometimes the suggestion that we should be wary of taking evolutionary theory as settled scientific fact, since many points remain contentious, and since quite often the motivation for the reinterpretation or even rejection of Genesis and the Bible is atheistic naturalism masquerading as evolution (and as science generally). We should be cautious, some argue, of according too much respect to what is a controversial scientific theory, a warning also aimed at contemporary theologians who too often accept unquestioningly this scientific starting point as the only legitimate gateway to their theological work.

2. **Critique of evolution as operating randomly:** This is a main complaint of many who reject theistic evolution (especially those in the ID camp)—they cannot accept the claim that the process of evolution is supposed to operate randomly; more accurately, they cannot accept that a random process can be reconciled with God's providential oversight. They propose both logical and scientific objections to evolution, which we will explore thoroughly below, but the main thrust of the argument of most ID theorists is that common ancestry and natural selection operating in a random way have not, in fact, been demonstrated. Moreover, these claims are not compatible with the evidence we see in biology itself, particularly in the makeup and operation of cells and DNA, and their structures. This argument has been advanced in its most sophisticated form by Stephen Meyer.[3] Intelligent design, Meyer argues, is a more plausible explanation

3. In addition to Meyer's *Signature in the Cell*, see his essays in Moreland et al., *Theistic Evolution*. Another approach to ID theory can be found in Behe, *Darwin's Black Box*; also, Johnson, *Darwin on Trial*.

for these features of biology than natural selection operating randomly in nature. So ID theorists are particularly harsh on theistic evolutionists for their general acceptance of this key, indeed currently defining, view of modern biology—that it operates largely by chance—and for embracing the stark implications of such a view (which some do only reluctantly, some try to dodge, and others seem to quite warmly welcome, as we will see). They are particularly critical of the implication that God would have left the emergence, nature, and hierarchy of species to chance, and argue that this is not compatible with Christian theism in any reasonable sense. So they reject the position that God guides evolution from the beginning without the need for any further direct intervention. This view does not work, they hold, because of the large element of chance that is supposed to be present in evolutionary processes.

So creationists/ID theorists propose that an evolutionary process, defined by chance, is not good science, and also that it is not compatible with sensible readings of Christian theology. They regard these two points as significant criticisms of theistic evolution. They are very skeptical—often scornful, like their atheistic counterparts—of the attempts of various theistic evolutionists both to reconcile the apparent randomness of evolution with God's design plan and also to reinterpret Christian theology, especially Scripture, to make it cohere with this reading of how nature operates, and how the human species came about. This critique also leads naturally to the third critique, which focuses on the position known as methodological naturalism.

3. **Critique of methodological naturalism:** Many creationist/ID scholars claim that one of the big failings of theistic evolutionists is that they accept, often in an unquestioning way, the thesis of methodological naturalism about the way science works and should work. Here, we need to highlight the very useful and much appealed to distinction between methodological and metaphysical naturalism, now a commonplace in the broader discussion.[4] Methodological naturalism is the convention that when we are doing science, only physical, testable explanations may be considered for any phenomena under investigation. For example, if we are using the James Webb telescope to study distant parts of the universe, we will analyze the data and the evidence we gather purely from a physical, testable perspective, appealing only to scientific laws, causes and effects, empirical evidence, and reasoning based on appeal to these

4. This distinction was first introduced by De Vries, in his "Naturalism in the Natural Sciences."

physical phenomena. We rule out, automatically, almost by definition, as it were, any other type of explanation, such as appeals to intelligence, divine causality, or teleology (which would have to be explained by appeal to an intelligence). A study of the human cell, or the human eye, serve as further examples. In considering the origin and structure of the eye, an example much discussed in this literature, we notice that it is so intricate and complex (an empirical observation) that it crosses our minds that it must be designed. However, according to the convention of methodological naturalism, we could not bring this conclusion into our scientific work (even if we thought it was a reasonable conclusion based on our empirical study of the eye).

This methodological approach is to be carefully distinguished, its proponents insist, from metaphysical naturalism, the view that only physical, testable explanations should be accepted *for any phenomena under investigation in any discipline*. This is a stronger thesis, suggesting that only scientific explanations can count as genuine attempts toward explaining any happenings or phenomena in the universe in the pursuit of new knowledge. In short, metaphysical naturalism is the thesis that science is the only way to knowledge; it further implies that any question that cannot be approached through science is not a proper question, or at least that it must be reframed as a scientific question, and so forth. Paul de Vries and many others have argued that this distinction is crucial in the religion/science debate. Moreover, it is a simple logical mistake to confuse the two, or to suggest that methodological naturalism implies metaphysical naturalism. And of course these thinkers pointed out, quite correctly, that the confusion between the two is rife in the literature, especially on the naturalistic side, and sometimes even creeps into textbooks, usually inadvertently.

Many thinkers have found this distinction helpful because it allows us to insist that science be confined to physical, testable explanations ("natural explanations") while denying that all of reality must be approached or understood this way. Putting the point differently, we *could* bring design into our explanations in the broader realm of the rational (identified in chapter 1, such as in the traditional argument from design for the existence of God). So it is a very useful, indeed we might say vital, distinction that many thinkers embrace. The correct way, perhaps, to think about this is to note that when we look at the human cell under an assumption of methodological naturalism, we would rule out any inferences to a designer. Yet, if the evidence suggests to us that design is present, we would, in proposing this conclusion, note that it is not being presented as a scientific conclusion, but as a *rational* one. We are not looking at different parts of reality under the auspices of science and rationality/philosophical reasoning, respectively, but looking *at the same parts of reality* under two different auspices. Obviously, we have to be careful with such moves and distinctions, but as I will show below, this is a helpful way to reason about some aspects of nature. However, ID theorists and creationists are wont to reject the convention of methodological naturalism because they believe it arbitrarily rules out the possibility of design, and that it often functions in practice as metaphysical naturalism, however inadvertently.

The Biblical Critique

Beginning then with the biblical critique, what are we to make of the argument that theistic evolution involves denying key historical claims in the biblical text, basically obliging us to choose between them? Since we cannot choose theistic evolution and remain biblical Christians, we must reject it, and thus evolution itself (or most of the theory). This is indeed a strong claim, especially since it puts the Bible on a collision course with a widely accepted scientific theory (albeit not one without controversy). Some creationists are quite direct about this, seemingly giving the biblical text an absolute priority and so finding any outside theory that appears to conflict with it as wrong on its face. Others are more circumspect, perhaps a bit reluctant to go against current science but also not wishing to undermine the Bible.

We should emphasize, firstly, that all traditional Christian theists in this broad discussion accept that the Bible is the revealed word of God, so there is concern about any theory that *appears* to question it in a fundamental way. Most theists would be initially cautious about a scientific theory that appeared to conflict with Scripture. The second point is that it seems obvious we cannot start with the premise that everything in the Bible must be read literally in every respect, that none of its expressions are intended as allegory, metaphor, story, or allusion. It would be going too far to suggest that the biblical writers, for instance, never employ what seem on the surface to be factual claims in order to convey in a metaphorical way deeper moral and spiritual truths. Even if one accepts biblical inerrancy—the doctrine that the Bible is without error—this still leaves the question as to what this or that passage, or this or that account or story or report of events and happenings, actually teaches, says is true, presents as historical fact. Even with this doctrine of Scripture (which we should note is also subject to various interpretations), there still remains disagreement concerning the general questions of biblical understanding, while acknowledging that this is not an intractable problem, at least in most cases. And many thinkers hold to a less strict standard of biblical infallibility, meaning that the text is without error in its fundamental claims about God, faith, salvation, and morality, but not necessarily with regard to every factual and historical detail.

Our third point then is that this is not a dispute about the doctrine of inerrancy or indeed concerning biblical infallibility. We don't have, on one side, the creationists claiming that because the Bible is inerrant, the creation story must be read as mainly presenting historical fact, and therefore the evolutionary account must be rejected, and the theistic evolutionists on the other claiming that since the Bible is not inerrant the Genesis story need not be read literally and might be reconciled with evolution. For the fact is that many Christians who are interested in the issues of this book accept some version of the doctrine of biblical inerrancy or of biblical infallibility. Despite agreement that the Bible is truthful and contains no fundamental error, there is still room for some divergence about which truths are being conveyed in various texts and narratives (as well as concerning their level of importance in the overall Christian message). The fact is that long before the theory of evolution came along, the Bible, although accepted by Christians as the revealed word of God, was nevertheless often interpreted as mostly metaphorical in certain areas, and one of these was with regard to the book of Genesis.[5]

One can see this clearly from a study of what is now a two-thousand-year history of interpretations and commentaries on the Book of Genesis, across a quite wide range of scholars and thinkers, as Stanley Jaki has illustrated.[6] These works range from Philo of Alexandria (c. 20 BC–c. AD 50) and the early Jewish commentators (some of whom showed early influence of Platonic and Aristotelian ideas, ideas that were to remain dominant until quite recently) as well as early church writers,

5. Although we are focusing here on the early Genesis, other parts also raised conflicts with science; for a discussion of the emerging conflict between the story of Noah's flood and progressive scientific discoveries, see Brown, "Noah's Flood." For a similar approach from the point of view of geology, see Greene, "Genesis and Geology Revisited."

6. See Jaki, *Genesis 1 Through the Ages*.

including Theophilus (second century), Irenaeus (c. 125–c. 202), Origen (c. 185–c. 253) (who suggested that the creation of heaven on the first day refers to the spiritual realm and not to a particular moment in time), Gregory of Nyssa (c. 335–c. 394), and others, several of whom adopted a rational approach toward scientific matters when trying to discern the correct understanding of the biblical text. One persistent worry for early Christian commentators on Genesis was how to reconcile the appearance of the sun and the moon on the fourth day with the creation of light on the first day. Another concerned the relationship between the firmament and the waters, especially the upper and lower waters, accompanied by reflections about the coming into being of the various lifeforms. Then came Augustine (354–430), a very significant thinker on this matter, but not one without apparent contradiction and an often inchoate approach to specifics. He worried a literalist interpretation that could not be reconciled with what we know from reason risked undermining the overall truth of the Bible. Augustine was one of those thinkers who began to struggle with the "how" of Genesis as well as the "what," suggesting, for example, that the firmament must be regarded as historically real, but that its nature is open to interpretation. There were many literalists along the way as well, including among the Protestant reformers, and this often led to fierce debate and a clash of views. Jaki's meticulous and detailed study of the great thinkers throughout the history of Christianity illustrates clearly a wide spectrum of views on how to read the creation story in Genesis. Jaki himself is extremely critical of any kind of concordism (the view that all scientific discoveries can be made to cohere with the Bible), and suggests that the results of his study show that Genesis is not to be read as a scientific text, but as conveying deeper philosophical and theological points.[7] Jean Pond is correct that Jaki's study shows that the history of the interpretation of Genesis is not encouraging when it comes to finding agreement among scholars and commentators![8]

Sometimes, as we have seen above with Jaki, it is suggested that we should handle possible conflicts between scientific theories and biblical accounts by simply declaring that the Bible is not a science text. Versions of this approach have been suggested by Howard Van Till, Fr. George Coyne, as well as perhaps by Pope John Paul II, and maybe even by St.

7. See Jaki, *Genesis 1 Through the Ages*, 259–90; also Carlson, *Science and Christianity*; also Craig, *In Quest of the Historical Adam*.

8. See Pond, "Mutual Humility," 93.

Augustine in his reflections on the topic.[9] This is also the favored interpretation of an argument that Galileo used in his defense, saying famously that the Bible tells us how to get to heaven and not how the heavens go.[10] Indeed, it is a familiar argument in this general discussion. Yet, we should be careful not to fall into what can become a too glib response. While it is true that the Bible is not a science text, this rejoinder still seems too simplistic for handling some of the deeper questions. The key question, as Grudem has noted, is whether or not the claims made in the Bible about significant factual matters, including those that might be classified in the domain of science, are true. That is the main question that must be faced. It is correct to point out that the Bible is not trying to teach science, that it is not proposing scientific theories, or offering scientific explanations of physical events, and so forth. Yet it does occasionally make claims that might be classified as scientific, or that might seem to contradict what we know from science, not to mention the huge number of straightforwardly factual claims it makes. So, for example, if the Bible records that Adam and Eve were the first humans, this could be classified as a scientific claim (even though no further scientific explanation is provided), but it is one that clashes with the theory of evolution, which denies this claim. So it is too facile to respond to some of our worries by simply saying that the Bible is not a science book, because it papers over these difficulties and leaves them unaddressed.

Grudem has provided a sophisticated defense of the creationist approach to the Bible. He argues that many claims of theistic evolution are incompatible with a fairly literal reading of Genesis, and he extends the argument to suggest that theistic evolution is irreconcilable with many other biblical themes. It will be instructive to discuss his view here to bring out the key differences between the view I wish to defend and a sophisticated creationist approach, which will also allow me to show where I think Grudem goes wrong or oversteps the mark. We should note, initially, that he is working with a problematic understanding of theistic evolution, which he defines as follows:

> God created matter and after that did not guide or intervene or act directly to cause any empirically detectable change in the

9. See Van Till, *Fourth Day*, 75–93; Coyne, "God's Chance Creation"; John Paul II, "Truth Cannot Contradict Truth"; Augustine, *On Genesis*.

10. See McCarthy, "Letter to the Grand Duchess Christina," 23.

natural behavior of matter until all living things had evolved by purely natural processes.[11]

The problem with this definition is not only that it is too vague, especially in its use of the term "natural," but that it runs together several versions of theistic evolution. For example, most theistic evolutionists hold that God designed the universe, including life and the evolutionary process overall, and that it is heading in a direction intended by him. This also seems to be the view of Francis Collins, for example, though not of Kenneth Miller, who suggests that God does not direct evolution.[12] Collins suggests that although God designs the process overall, he does not later intervene *directly* (as ID theorists claim) to further guide or direct any of the naturally occurring processes on earth, including the process of evolution. But if we take the phrase "God created matter" in Grudem's definition to include the idea that the matter has a type of built-in guidance plan (this includes obeying the scientific laws that God has put in place), which is the way it must be understood, then it would not be true to say that God created matter but does not guide it later on (which is what the definition says, although it would be true to say that God does not *intervene directly* in nature later on). This point will be important in our discussion of whether nature operates deterministically or indeterministically, because, if the former, one can conclude that God is directing it from his initial set-up, and that this is *empirically detectable*, to some extent, thereby challenging Grudem's definition.

Grudem identifies twelve points where (theistic) evolution differs with the biblical account of creation taken as an historical narrative. These include the historical claims that there was a first pair of humans (Adam and Eve), that God acted directly or specially to create Adam, that God created Eve out of Adam's rib, that Adam and Eve were initially sinless, that man fell from God's grace in an act of disobedience, that God created the species complete and intact, that all human beings are descended from this original pair, and so forth. Given this, the theory of evolution, which contradicts all of these claims, cannot be true. Even if one were to read some of

11. See Grudem, "Theistic Evolution Undermines Twelve Creation Events," 784; also Grudem, "Incompatibility of Theistic Evolution," 67–68. Grudem notes this is the definition the editors of the volume in which his essays appear (Moreland et al., *Theistic Evolution*) have chosen as representative of their understanding of theistic evolution. Though, some contributors, such as Stephen Meyer, have a more nuanced view, see Meyer, "Scientific and Philosophical Introduction," 33–40.

12. See Miller, *Finding Darwin's God*, 210–12.

these claims as symbolic or poetic, the tension between evolution and the plain meaning of Genesis remains insurmountable.[13] He also denies that the biblical text suggests a metaphorical or allegorical reading, arguing that this reading "depends upon a prior commitment to an evolutionary framework," which he believes ID arguments show is unjustified (so the scientific case against evolution seems essential to his view).[14] He also points out that the historicity of several events in Genesis 1–3 is affirmed by Jesus and in other places in the New Testament. So if these events did not take place, then Jesus and the New Testament writers are in error. A denial of the historicity of these claims would also undermine a number of important Christian doctrines that rely on them, such as the atonement, the resurrection, and even the equality of human beings.[15] Grudem further holds that if we reinterpret Genesis to accommodate the theory of evolution, and specifically a theory of theistic evolution, this would result in undermining the reliability of the Bible as a whole: "When significant historical records in Scripture are explained as not being truthful records of actual events, then eventually other passages of Scripture—usually those unpopular in modern culture at the moment—will eventually also be explained away as untrustworthy. . . . [M]any . . . followers of those who hold to theistic evolution today will abandon belief in the Bible altogether, and will abandon the Christian faith."[16]

Although Grudem's essay is thoughtful, scholarly, careful, and thorough, and he makes probably the best case one can make for his side of the debate, I don't find his view persuasive. Let us first consider the last point noted about whether reinterpreting the Genesis story to make it compatible with evolution undermines the Bible as a whole, one of Grudem's two main criticisms of theistic evolution. Is it correct to insist that if we accept evolution, then we can't really avoid undermining the Bible overall, since if one part of it is open to a metaphorical or symbolic reading, then so are other crucial passages, relating to doctrines such as the incarnation, the resurrection, and so forth? To see why this argument is not persuasive, we might invoke a comparison with Christian teaching on sexual morality. The history of Christianity has affirmed clear teachings on sexual matters, based on the Bible, the theology of the early church fathers, and

13. See Grudem, "Incompatibility of Theistic Evolution," 73.

14. See Grudem, "Theistic Evolution Undermines Twelve Creation Events," 786.

15. See Grudem, "Theistic Evolution Undermines Twelve Creation Events," 835–36.

16. Grudem, "Theistic Evolution Undermines Twelve Creation Events," 822.

on the work of philosophers in the medieval tradition. These teachings have been strongly challenged by the cultural sexual revolution that has occurred gradually over recent decades, quite a bit of which was inspired by atheist and secularist views (themselves influenced by Enlightenment liberalism). As a result, modern practices have significantly challenged the allegiance to Christian teachings on sexuality in numerous cultures, with many now regarding these teachings as not only outdated but even as morally wrong. Even some churches are beginning to adopt this perspective, or at least acquiesce to it in the face of not having much choice.

So how should theologians, religious clergy, churches, and philosophers of religion think about this significant cultural change? Specifically, how could they go about *reconciling* biblical teaching with modern Enlightenment views on sexuality in a way that is plausible? The simple answer is that this is very difficult to do. It seems they have two options. First, they could argue that the Bible was wrong all along on sexual morality (but right on many other topics), that the biblical writers were influenced by their time and place, that they allowed their own cultural values too much sway over their understanding of human sexuality. Today, however, we are more enlightened, and we have progressed to the correct understanding. And it must also be true, of course, that this modern understanding of sexuality is the view intended by God in his creation. A second option is to argue that we have just misread and misinterpreted the Bible on the topic of sexual morality; that it actually teaches an Enlightenment liberal approach beneath the surface, and it has just taken a long time to bring out this meaning. This second view might seem like a stretch, but one might suggest perhaps that the texts were distorted along the way, or that they are incomplete, or that they must be read in conjunction with other aspects of human knowledge and discovery, in an attempt to produce a plausible rationale to explain why they actually mean the opposite of what they appear to mean. (A third option, of course, is to retain the traditional meaning and to reject the Enlightenment view, a move that would preserve consistency and would not threaten the text.) Now it seems that both of these moves *would* undermine the biblical texts *overall*, since they lack credibility. The first option seems to go against inerrancy and the factual accuracy of the Bible with regard to central teachings about human behavior. It is also hard to avoid the conclusion that the second option is an attempt to re-read back into the Bible views on sexuality that one favors but which the Bible rejects in order to make one's religious teachings consonant with contemporary cultural practices.

However, these problems are not applicable to the creation story. This is because, first, unlike with sexual morality, there was *no agreement* early on about whether the Book of Genesis and other relevant passages were to be taken as straightforward literal historical fact. There are many different interpretations of how to read Genesis throughout church history across a wide range of great thinkers, as Jaki's detailed textual study shows. Therefore, second, the claim that it is easy to misread the Bible on the issue of the meaning of the creation story is much more plausible than it is for the topic of sexuality. Indeed, it might be more accurate to say that the progression of scientific discovery, including with regard to the evidence for evolution, eventually gets us to the point where it helps us to fill out more accurately the story of creation. A third observation is that once we accept that the text wishes to convey deeper truths using familiar images and metaphors then it is not correct to conclude that the Bible is "wrong" in its account of creation. What we should more accurately report is that it is only the surface literal details that seem to contradict modern science, but we don't take such details to be the actual way God created the universe and life.

So it seems that a metaphorical or poetic reading of Genesis does not undermine the Bible in the way that the rejection of its teaching on sexual matters does. Nor would scientific discovery undermine the Bible overall. One might think this might occur because one could argue that if the creation story is undermined by science, then why could the story of the resurrection of Jesus not be undermined in a similar way, because our scientific knowledge precludes anyone coming back from death (this is often what the atheist argues). However, this argument does not work because it is not quite correct to say that science undermines the creation story if this means that science shows us that miracles cannot occur, or that God cannot intervene in creation. The creation story is only undermined in the weak sense that the literal details of it might reflect a metaphorical way of illustrating deeper truths. Let us not forget that such deeper truths are at least fivefold: that there is only one God; that God created the universe; that God created life; that God created human beings; that men and women have the same nature; and that God intended for human beings to be at the top of creation.[17] These truths

17. See Jaki, *Genesis 1 Through the Ages*, 259–90, for a discussion of this point.

were accepted by almost all theologians and biblical scholars, who otherwise disagreed among themselves about the literal details of the Genesis account, including about the origin of the first humans. Moreover, these truths imply God's *miraculous* intervention in his creation in some form, so miraculous divine action is not undermined by our rereading of the creation story. Indeed, it is affirmed by it. And so the creation story is not undermined in the sense that it is rejected as false. The same reasoning applies to the resurrection and other important doctrines that involve divine miraculous activity.

There are two strong additional reasons to reject Grudem's argument that the whole of the Bible is undermined if we pursue the metaphorical or poetical strategy (an argument also advanced by atheistic naturalists, but it fails for them too). The first is that, taking into account what we can describe as expert, responsible biblical exegesis, we know that one of the foremost Christian beliefs, the resurrection of Jesus, is presented in Scripture *as a historical event.*[18] The narrative of the resurrection is not presented as a metaphor or motif or allegory to convey a deeper truth about (perhaps) the exemplary spiritual and moral nature of Jesus Christ. Therefore, any interpretation that suggests it is intended to be metaphorical, or even that it is best read as a metaphor, is not being faithful to the text. This position does not settle the question of whether the resurrection happened. We don't need to do that here; we only note *what is being claimed* for the resurrection in the text; the plain, commonsense meaning, according to responsible scholarship, is that it reports that Jesus rose from the dead. So any thinker (such as John Haught, Arthur Peacocke, or philosopher John Hick) who denies this historical fact or reinterprets this "event" as intending to mean something non-literal is incorrectly reinterpreting the Bible to make it more compatible with modern science. However, this kind of argument does not seem to apply to the Book of Genesis because not only does the account carry resonances of a poetical or metaphorical reading, the plain meaning of the creation story is disputed by responsible biblical scholarship and *has been since the early church.* We have seen this clearly in the various attempts to understand Genesis throughout the history of biblical scholarship. So the Genesis account of creation is not in the same category as the narrative of the resurrection of Jesus, and therefore the argument that interpreting the creation narrative metaphorically undermines the reliability of all biblical

18. The kind of responsible work to be found in Wright, *Resurrection of the Son of God*; Johnson, *Real Jesus.*

texts is very weak, and must be regarded as overreach. Nor, I believe, are doctrines such as the atonement or even original sin undermined by a metaphorical reading, since the underlying truth is that humanity is deeply flawed and given to sin, even if the details of how we got into that state are poetically conveyed in Genesis.

The second reason is that we have strong scientific evidence for at least some parts of the theory of evolution, and so if we are to be responsible thinkers, we must take this into account when understanding Genesis. And the fact is that this evidence gives us further reason to believe that the readings suggested by such thinkers as St. Gregory and St. Augustine are correct (imprecise though they are). It is very important to keep in mind that these readings were common and prominently discussed long before the theory of evolution (so they were *not* introduced simply as a way of dealing with an objection raised by the theory). So Grudem is incorrect to claim that the reinterpretation of Genesis is motivated by "a prior commitment to an evolutionary framework of interpretation."[19] As we have seen, he himself rejects evolution in part because he is persuaded by the arguments of the intelligent design theorists. This is where the scientific consensus around the evidence for evolution is important because it should give any responsible thinker pause. It is true that some aspects of evolution are in dispute (both within and outside biology), and I agree that we must be mindful of this, wary of the zeitgeist and the antireligious views often associated with the theory in thinkers like Dawkins and others, which often seem to pervade large parts of the scholarly exchanges. Those who reject the Bible because of evolution have a rather poor argument; sometimes they may be rationalizing a prior commitment to atheism, like so many do today, because, like Thomas Nagel, they would prefer it if the religious view of reality was not true.[20] This is a human failing, but not one that serves to undermine the biblical text.

More generally, I think we have to conclude that when it comes to the question of biblical interpretation, we run into what seems to be an intractable problem. For it seems that we cannot settle definitively on an "accurate" and "correct" interpretation of important passages in the text of the Bible (for our purposes, we will confine ourselves to Genesis). Respected, brilliant scholars throughout history have disagreed about this matter over and over again, despite Grudem's arguments. Indeed, he

19. Grudem, "Theistic Evolution Undermines Twelve Creation Events," 86.

20. See Nagel, *Last Word*, 130.

does not refer to the various interpretations of Genesis that have been offered from earliest times, many of them inspired by attempts to reconcile the biblical account with (scientific) details from the natural world. This problem of interpretation, a vexing one for Christianity, is often cited by critics who reject the Bible altogether as unreliable. It is also one of the reasons that so many Christians, especially today, blatantly reinterpret biblical passages or ignore them if they do not cohere with contemporary cultural beliefs and values they wish to uphold. However, to be both logical and responsible, we must reject the view that anything goes, that there are no restrictions on how we might read the biblical text. That would be a ridiculous position, a reckless submission to intellectual laziness, and one inspired by contemporary fads and ideological biases. I agree that we must be guided by the text, by the study of historical events, by archeology, by the study of languages, by theological and philosophical analysis, when working with the biblical text. We can describe an approach that takes into account all of these areas as an exemplification of expert, responsible biblical scholarship. Moreover, we must be guided by the plain meaning of the text when read reasonably and with common sense, taking into account the areas of study mentioned. We might use an analogy with the question of how to interpret civil law. We will rightly reject any argument that suggests that anything goes, yet there is still room for interpretation, even taking into account the plain meaning of the text, its historical origin, legal precedent, and its application to current circumstances.

Grudem often turns to specific points of interpretation with regard to Genesis to strengthen his argument. He thinks, for instance, that Genesis 1:26–28, where God says "be fruitful and multiply and fill the earth and subdue it, and have dominion over the fish of the sea and over the birds of the heavens and over every living thing that moves on the earth," must be understood as historical narrative.[21] His reasons are: (1) this passage occurs in the first chapter of the Bible; it is dealing with a historical question—an explanation of how things came into being; (2) the chapter moves sequentially through creation—light, land and sea, plants, the heavenly bodies, fish, birds, animals, human beings; (3) there is nothing in the passage to suggest that it is not meant as historical literature. While Grudem's view is not obviously incorrect, I think he exaggerates. For we can agree with all his points, but still hold that it is intended as a

21. See Grudem, "Theistic Evolution Undermines Twelve Creation Events," 789.

metaphorical account of the underlying facts that God created the universe and life, with human beings purposefully at the top of the evolutionary tree. Is the latter view plausible? I think it is, given the structure and rhythm of Genesis, which does have echoes of poetry or metaphor, and also given what we now know from science. This latter point is important because it does give us an additional reason as support for a metaphorical reading. Grudem's three points are, I believe, all equally true of a metaphorical reading.

Another example is his dispute with John Walton about what the claim that God "formed man of dust from the ground" means in Genesis, a claim that is inconsistent with evolution. Following John Currid, Grudem argues that this phrase should be understood as a factual, literal claim, while Walton suggests it is best read as a metaphor for Adam's mortality, arguing that the verb for "formed" need not refer to forming a material object. Walton supports his view by appeal to Psalm 103:14, where the word "dust" is used in the context of mortality.[22] Grudem believes that Walton does not pay sufficient attention to the context of Genesis, because Genesis 2 provides a more detailed explanation of how God created human beings. Walton's mistake, he thinks, is to make a verse about the creation of man into a verse predicting man's death. While this kind of detailed textual exegesis is useful and surely has its place, it is hard to see that it can settle such big issues. This is why I think it is best to interpret the whole of the Genesis account in a metaphorical way, an account that is making the deeper points mentioned. This is justified for two reasons: (1) a reading of the text does suggest a poetical or metaphorical reading based on the form and cadence of the language, as well as the subject matter (though this by itself is not a definitive reason), which is why this was one of the ways the text was read from the earliest days of biblical scholarship, and (2) our factual discoveries (including with regard to evolution) progressively over time (many of which would now be classified as part of science, of course) reinforce this metaphorical reading of Genesis, since a literal reading cannot be reconciled with many factual claims about nature (not just with regard to evolution). Indeed, as Jaki shows, this was one of the reasons that early thinkers in the church suggested that Genesis was likely metaphorical.[23]

22. See Grudem, "Theistic Evolution Undermines Twelve Creation Events," 799–800.

23. See Jaki, *Genesis 1 Through the Ages*, 65–98.

So I think that both approaches—that of Grudem and Walton—are on the wrong track. It is best not to haggle over the details of the biblical account in an attempt to seek out the "correct" understanding of this or that phrase or claim (though this work is not without value). It is more reasonable to read the whole narrative as a storied account employing metaphors to convey the deeper truths concerning God's creation. It is less pressing then to identify which particular truths are intended by various parts of the overall metaphorical account. However, there are some textual details identified by biblical scholars, including Grudem, that should give us pause and that are not so easily dealt with. One example he points to is that 1 Chronicles 1:1 refers to the genealogy of the human race, tracing it back to Adam, a view also it seems held by Luke and Paul in the New Testament, though it is not entirely clear that they were referring to actual historical events. (Paul in 1 Corinthians 15:45 says that "*It is written* that the first man Adam became a living being.")[24] And this is the problem—once a passage is not crystal clear, it leaves open a pathway to reconcile it with a metaphorical reading of Genesis.

But leaving all this aside and trying to be as neutral as possible, being neither prejudiced against the theory for religious reasons, nor exaggerating the evidence for evolution (being dispassionate toward the theory), and also open to narrative readings of the Bible where justified, we can say that the theory does seem to reinforce the largely metaphorical reading of Genesis that was developed in various ways by many early scholars, long before the idea of evolution was an issue. Earlier scientific work also raised the same questions about how to read Genesis, so this issue is not unique to the theory of evolution. The early thinkers of the church struggled mightily in trying to reconcile their own (what we would now call) scientific observations with the biblical account, and none of them were motivated by a prior commitment to any evolutionary framework. Indeed, if the theory of evolution comes to be rejected in the future, we will likely return to the early range of disputed interpretations about Genesis, and not to the view that it should be read as making a series of true historical claims, as Grudem argues.

24. See Grudem, "Theistic Evolution Undermines Twelve Creation Events," 791 and 798 (my emphasis). Paul seems to be bringing up Adam here as the first man in order to contrast him with the "last" Adam, Christ. He does not appear to be speaking of historical events (see vv. 46–49). (I owe this point to Bill Stancil.)

The Chance Critique

We now turn to consider the critique, particularly from ID theory, that theistic evolution too readily accepts the claim that evolution operates randomly; moreover, that this is not compatible with the view that God created not only life, but the universe, for a particular purpose. Intuitively, this is a sound criticism to raise since it is founded on what seems to be the most reasonable approach to Christian theology. That is to say, it was an essential part of Christian theology from the earliest times—a bedrock belief, if you will—that God's providential wishes influence his creation. The universe and life follow a (design) plan that is laid out by God; there is an overall purpose to the nature and structure of life and the universe. It does not seem compatible with Christian theology, at least in a general sense, to say that there is no purpose to creation, no overall meaning to life, no plan or direction to the universe. (We will come back to the interesting topic of human free will later.) Logically, God could have created a universe devoid of meaning and purpose, but this is extremely unlikely, and it is also inconsistent with the biblical texts and with centuries of Christian theology. So this is one reason why evolution, when it is presented as operating largely by chance—a process responsible for bringing into being all of the species, including us, with all of our particular features—seems to undermine purpose in general, and raise a challenge to God's design.

So we must say initially that ID theorists are raising a very plausible criticism of theistic evolution with regard to the vital matter of design and chance. It is not one, moreover, that is aimed only at the theistic side in the debate; it also applies to the atheistic critique of religion. One of the key reasons thinkers like Dawkins and Dennett harp so much on the vital role chance plays in the way nature unfolds is (as we noted in chapter 1) so that they can promote the larger ideological point that the absence of purpose is further evidence against a designer. So both creationists/ID theorists and evolutionary naturalists seem to agree on the point that an absence of design is an argument against theism. Since I am arguing in this book that evolution does *not* operate randomly, I am in fundamental agreement about the problematic nature of accepting that evolution operates with a large element of chance. But there is much more to be said, of course. One problem with modern biology, and it is a very large problem, is that it teaches and explains the theory of evolution as operating randomly throughout nature. This is a very large part of how the theory is explained and taught, and indeed the supposed absence of design in how evolution works has almost become a dogma attached to the theory (in the sense that it is a belief that cannot be questioned). The idea does go back to Darwin, of course, who often commented on the random way that evolution seems to operate; he drew the implication that this suggests the species that emerge from the process do so randomly, along with their various properties and characteristics. An emphasis on this random feature has generated a significant tension in the theory itself, one that Darwin acknowledged quite well, but that his contemporary disciples often will not concede.[25]

The tension is generated by the fact that the intriguing complexity we see in species, especially *Homo sapiens*, which suggests not only design in nature but overall meaning and purpose in life, conflicts with the view that chance and randomness pervade how nature evolves. The complexity of humanity in particular, but also the general complexity of nature, would give any reasonable person pause on the question of randomness, except perhaps those with ideological commitments against the possibility of design (ideological commitments that have no place in pure science). Moreover, the fact that physics is *not* taught this way—as operating with a large element of randomness—should also give influential evolutionary biologists pause. Physics is taught as operating deterministically (especially at the practical level), and without any elements of chance, yet biologists have not considered how it is that one area of physical reality seems to operate in a deterministic way, and the other, which it is usually argued is made up of the same sort of physical stuff, does not. The more general question of whether there is any chance in nature overall remains an *open question*, as I noted in chapter 1. So, ID theorists are right that there is a serious question to be asked about the place of chance and randomness in biology.

ID theorists appear to rule out the idea of chance having much or any role in nature, and they wish to claim, therefore, that since nature provides evidence of design, this means that there must be a designer. They also hold that an intelligent designer *directly* caused certain outcomes in nature, especially in evolution, that contribute to design and purpose, rather than saying that these outcomes come to be from natural causes, under the general direction of God. (This latter view is the mainstream view of how God guides evolution, and is also held by many

25. See Darwin, *Origin of Species*, ch. 6.

theistic evolutionists, who then face the problem of how to reconcile God's guidance with the large element of chance they believe exists in the evolutionary process, and sometimes in nature in general.) ID theorists go further, insisting, crucially, that the inference to an intelligent designer is a scientific one because it was arrived at *empirically*. It is these two claims and their rejection also of the view that God could not have built design into the universe from the beginning, and so theistic evolution is on the wrong track, that I take issue with. They often use theistic evolutionist Francis Collins and other thinkers in the BioLogos group as their whipping boys! Several other theorists are often accused of holding similar views, such as Fr. George Coyne, John Haught, Denis Venema, and Howard van Till, though there are important differences between their respective positions (as we will see when we discuss the work of several of these thinkers in chapters 5 and 6). But the general worry about the approach of these thinkers is that they are unable to reconcile in a satisfactory way evolution operating by chance with a belief in God's providence.

Perhaps the ID critique of these thinkers is a little unfair. I think we have to acknowledge that Collins, for instance, is genuinely grappling with a difficult problem, and is making an honest attempt (for a non-theologian) at working out how to put evolution and his Christian theism together in a way that makes sense. There is no doubt that his view is vague (as we will see in chapter 6), and that the difficulties seem especially acute in those areas where he is trying to reconcile randomness in evolution with God's providence. Collins settles on one approach from a range of possibilities that are available to theistic evolutionists, though it is quite hard to get clear about his position because of the vague way he develops it. We see some of the same difficulties in other theistic evolutionists whose work we will come back to later.

One important thinker who has developed the ID view in a sophisticated way is Stephen Meyer. Thinkers like Meyer generally accept some parts of evolution (particularly microevolution, some parts of natural selection, and other concepts) but they reject the idea that the process operates randomly. Showing sophisticated knowledge of both the science and the theological issues, Meyer defends ID theory over various forms of theistic evolution. He is not simply observing that, in general, evolutionary patterns are too complex to have come about by chance, but is trying to illustrate this point in a more specific way by means of an analysis of the make-up of DNA. He has argued that views like those of Collins are close to being contradictory because they seem to say that God is directing a process that in fact operates by chance. However, Meyer is aware that some theistic evolutionists do not hold such a vague position. Some, like Denis Lamoureux, hold that God is directing evolution in a more deterministic way than Collins seems to allow for, and Meyer takes on that particular view in an attempt to illustrate that ID theory is a better explanation, one more in line with the evidence. This seems to be part of his general argument that ID theory is best supported by the evidence; it is not one crafted either to deny the theory of evolution or to respond to the objection about it being influenced by chance. Most ID theorists also retain a traditional, fairly historical reading of Genesis.

Lamoureux (who describes his view as "evolutionary creation") appeals to several metaphors to try to illustrate how he thinks evolution works, including the example of the development of the human embryo.[26] The human embryo contains everything in it from the beginning to drive its development into an adult organism; nothing new is added along the way that is not contained potentially in the initial make-up. Consequently, the final state of the embryo is determined by its nature at the beginning. Evolution, Lamoureux contends, works in the same way. God has "programmed" the process into the ingredients of the universe from the beginning, together with the laws of physics, and so the progress of evolution, along with the species that eventually develop, including *Homo sapiens*, had to emerge in a deterministic way. Also calling this view "teleological evolution" to reflect the fact that it has built-in purpose, Lamoureux's position also serves as a critique of ID because there is no additional need, then, for God to intervene. Lamoureux further objects to ID theory because it involves what he describes as a "violation" of natural laws since it requires God's occasional intervention in nature; in addition, it sounds too much like a "God of the gaps" explanation in the sense that when we find an event or process in evolutionary development that has no current scientific explanation (namely, complexity in organs such as the eye and also in the structure of DNA) we invoke God's direct intervention to explain it. Lamoureux holds that "teleological evolution" explains these effects since the ingredients for their eventual unfolding are built in by God from the beginning. (He provides no discussion of the roles of chance and randomness in evolution, so it is difficult to know how he would go about

26. For Lamoureux's view, see *Evolutionary Creation*; also his essay (to which Meyer mostly refers), "Evolutionary Creation." See Meyer's essay, "Difference It Doesn't Make," especially 229–36.

reconciling chance with the teleological position he defends in his work, a serious omission in any account of theistic evolution.)

Meyer objects to Lamoureux's view by first suggesting that we should not rule out (almost as a matter of policy) God's possible intervention along the way in nature, since this matter should be determined by the empirical evidence before us. Another reason in support of God's intervention, and the heart of Meyer's argument, is his claim that evolutionary mechanisms are not sufficient to explain the origin of living forms.[27] So he is not impressed with the argument, often offered by Lamoureux and others, that we don't need to bring in a "god of the gaps" because we already have an evolutionary explanation for biological complexity (or are likely to have one in the future). As ID theorists often insist, what we have are "just so" stories that ignore or downplay the difficult problems at the practical level (a well-known, and apparently intractable, problem facing the evidence for natural selection).[28] Moreover, this complexity suggests design, lending further support to the ID argument.

An interesting third argument advanced by Meyer, directed against Lamoureux's approach in particular (which he describes as a "front-loaded" approach), is that this view appears to be problematic for *scientific* reasons.[29] A key issue for Meyer is the distinction between inanimate matter and animate matter, or living things, which eventually come into being approximately twelve to fourteen billion years after the Big Bang (about two billion years into Earth's five-billion-year history), according to current estimates. He notes that Lamoureux is quite vague on what he thinks was present in the ingredients of the Big Bang that led to the eventual arrival of life, and he also does not discuss the various (very speculative) theories concerning the origin of life. He works with vague,

27. See Meyer, in Moreland et al., *Theistic Evolution*, 222–31.

28. For appeal to some of these "just so" stories, along with empty statements presented without evidence concerning the process of natural selection, see Dawkins, *Blind Watchmaker*, 77–109, 136, 240; also his *Climbing Mount Improbable*; Ayala, *Darwin's Gift to Science and Religion*, 65, 72. See my *Evolution, Chance, and God*, 50–59, for a discussion of this familiar problem with regard to the evidence for natural selection.

29. The intelligent design theorists appear to be ambivalent toward a front-loaded approach. Although Meyer critiques it here, Behe seems to favor it; see his *Edge of Evolution*, 165–66. George suggests that Behe has modified his position; see "What Would Aquinas Say About Intelligent Design?," 676–700. It is difficult for ID to accept a front-loaded view because it seems to defeat their project that direct divine intervention is required to bring about design in nature. A front-loaded understanding seems to be indistinguishable from a more traditional approach that God is guiding evolution (through secondary causes). We will return to these issues later.

abstract claims about how God could and likely did "front-load" everything in from the beginning, a vagueness that is typical of this literature, according to Meyer, and an approach that also, crucially, fails to come to terms with what the science shows.

Meyer believes that Lamoureux is vague on how the front-loaded view actually works. He argues that at times Lamoureux talks as if the laws themselves, which are present from the beginning of the universe, might be generating the new information that emerges later on, particularly leading to the origin of complex species like *Homo sapiens*. This is a very difficult matter to be accurate about, however, but Meyer suggests *that there is information present later on that was not present at the beginning*, and that "the chemical subunits of DNA lack the self-organizational properties necessary to produce the informational sequencing of DNA."[30] The front-loaded view denies this without saying, however, at which points novel developments emerge or a progressive move in evolution (or even in geology) is evident, except to say in general that whatever emerges along the way is built in from the beginning. Lamoureux and others are fuzzy also on whether it is the laws that are generating the novel developments or whether they are latent in the ingredients themselves and simply come out at the appropriate time due to the operation of the laws. Meyer seems to think that there are different levels of complexity in nature, and that the order produced by the laws cannot account for these deeper levels. (Though he notes that some scientists have suggested that there may be as yet undiscovered laws that govern life, a proposal that Meyer is skeptical about.)[31] These, however, are difficult matters to make a judgment about. Meyer doubts that the laws themselves could be responsible for the coming into being of the novel developments with new material or new information. He turns to the deeper question as to whether the initial ingredients could produce the new information necessary to move from simple life forms to more complex forms, one of the main claims of Lamoureux's front-loaded approach.

One of the main questions, according to Meyer, is whether "the information necessary to produce a functional gene (information-rich DNA molecule) [was] present in the arrangement of elementary particles just after the beginning of the universe?"[32] He argues that the answer is

30. Meyer, "Difference It Doesn't Make," 234.

31. See Meyer, "Difference It Doesn't Make," 226–28.

32. Meyer, "Difference It Doesn't Make," 229.

no because, in the first place, the biologically relevant chemical subunits of DNA themselves do not contain the information necessary for producing a functional gene or the functional information DNA contains. In addition, "there is no law that describes how these subunits self-organize into functional genes."[33] Meyer thinks this is evident from a study of the structure of DNA itself, and uses a helpful analogy: "to say otherwise is like saying that the law-like forces of chemical attraction governing ink on this page are responsible for the sequential arrangement of the letters that give this book meaning."[34] He goes on to suggest that if the chemical subunits of DNA do not contain this information, then it is hard to see how the simpler elements at the beginning of the universe could.

It is difficult to make a judgment about these logical and scientific claims, but I am not convinced by Meyer's analysis, challenging and probing though it is. It seems to me that God could have front-loaded the creation in the way Lamoureux suggests, in the way a computer program can contain at the beginning all of the subsequent functions that make themselves manifest in the running of the program over time. Is it a problem if Meyer is correct that there is information in the gene that is not present in the subunits of DNA that lead to the gene? He suggests that intelligent design provides the best explanation for the addition of the information necessary to produce the first living cell; that is, that an intelligence intervened directly at some point in the process and introduced new material and a causal process that led eventually to a functioning gene. (He argues more fully for this conclusion in *The Signature of the Cell.*) The theist, it seems, has to consider this possibility since God has the power to do this, and as Meyer notes, we should not rule out such a view by definition, or for "aesthetic reasons" (however these are defined). Nevertheless, it seems that God in his omnipotence could have done what Lamoureux suggests—built into the process the mechanisms, laws, ingredients, causal steps, and so forth that lead from (assuming Meyer is correct) subunits without the information to cells with the information. Does the fact that we can examine the subunits of DNA and note the absence of information and then examine the functional genes and discover that they now include new information mean that God could not have brought this about in the way Lamoureux suggests, or, less strongly

33. Meyer, "Difference It Doesn't Make," 230.

34. Meyer, "Difference It Doesn't Make," 233.

perhaps, that although God might have done it this way, he likely did not. I don't see that it is.

We need to consider whether the laws, together with the initial ingredients, along with an unknown element, could be built in by God at the beginning to eventually produce all the effects we see in the physical world. The "unknown element" is an element included in the ingredients that we either *cannot* detect using science, or that we have *not yet* detected. This is not necessarily an empirical issue, which is the way we usually think about it. If we cannot detect the mechanism, we are inclined to think it may not be present, or commit to a position of methodological naturalism that every process must be detectable by science (at least in principle) and postpone permanently the conclusion that nature in some of its aspects might be operating in a way that we cannot detect/solve. However, perhaps we can reframe this argument as a logical argument (rather than a scientific one) by suggesting that we can deduce that this is the way nature operates by thinking about nature as a whole, and not because we have any direct evidence for it in specific mechanisms. This is how the argument goes: It is more reasonable that the origin and development of life and later species were directed, given their complexity. No other view is plausible. But we cannot see any direct intervention, or empirical evidence for direct intervention, and since there are gaps in the steps that we cannot explain (such as from subunits to cells, according to Meyer) we conclude that the design is built-in or "front-loaded" (similar to the way it is in a computer program). It does not seem adequate to conclude that the gaps themselves are evidence of direct intervention (as we have seen in our discussion of Meyer's view). Is this view better than ID? I believe so, because it seems more in line with the evidence from science, even though the ID view is possible. (We must also be very careful not to be guided by current scientific dogma that evolution operates randomly, even though we accept methodological naturalism.) So the front-loaded view seems to me quite a plausible view. It is also consistent with our general understanding of how science works (though this is controversial, and we will return to this key point in the next section).

So ID theorists, despite Meyer's good work, are wrong that some form of the front-loaded view cannot succeed. Indeed, my own view in this book will be a form of the front-loaded view. The traditional and well-established distinction between primary and secondary causation (which we will explain further in chapter 5) relies in part on a front-loaded understanding of the operations of the universe. Perhaps in the future we will be able to detect more (or indeed all) of the currently missing steps in the progression of nature, or perhaps not, since God may have made some aspects of the process undetectable by the human mind (that is, not accessible to scientific discovery). If so, the nature of DNA would be another mystery that we would add to our list of things that seem beyond our understanding and that are so complex that they suggest an initial designer. So it doesn't seem to follow that God likely intervened directly to bring about the structure of DNA, as Meyer and the ID theorists argue.

Meyer's argument strikes me as being too vague to be decisive. Although his view is interesting and possible, I am not convinced that he has shown that (at least some form of) the front-loaded view is likely incorrect. However, the ID theorists are probably right that the view of theistic evolution that accepts that nature operates with a large element of chance cannot be defended (though we will consider a contrary view proposed by Arthur Peacocke in chapter 5). Meyer does point out that some thinkers claim that the pathway from subunits to DNA could have occurred by chance. Be that as it may, I am arguing in this book that there is no chance operating in the universe, so attempts to explain DNA as arising from a chance process are unnecessary. But even if chance were present in the universe, such an explanation would not be a *believable* explanation. It would be implausible to hold that such complex features arose out of a chance process, and it would also leave unaddressed what seems to be an insurmountable difficulty for a theist—that there are physical characteristics actually present that, just by chance, must develop in a certain way (because of God's intention). Reconciling chance and purpose is, as we will see, a particularly vexing problem for those theistic evolutionists who wish to affirm both in the universe. Although I will argue that some version of what Meyer has called the front-loaded approach is correct, it is very noteworthy that Lamoureux omits any discussion of the roles of chance and randomness in the process. Indeed, he glosses over this key difficulty, especially the question of whether he accepts that evolution operates according to chance, and how he would reconcile this with the deterministic nature of a front-loaded approach.[35]

35. Lamoureux spends much time on *one* of the problems facing his view—that it goes against the book of Genesis—but ignores the other large problem regarding the significant elements of chance that are claimed to be present in the operation of evolution; see his *Evolutionary Creation*, 241–82. There is no discussion of the concepts of chance and randomness in the book, and these terms do not appear in the glossary or index, nor do determinism or indeterminism. Lamoureux does devote a brief section

Critique of Methodological Naturalism

All of this brings us to the important topic of methodological naturalism. We need to consider the role it plays in the ID/creationist rejection of theistic evolution. The above critics usually claim that accepting methodological naturalism leads to various problems: It excludes ID from science—ruling it out by definition—even though ID is a reasonable view; it borders on metaphysical naturalism because philosophically it seems incompatible with non-natural explanations; it invites a too easy acceptance of the theory of evolution, including the controversial claim that the process is dominated by chance. Stephen Dilley has gone further and argued that an acceptance of methodological naturalism undermines most forms of theistic evolution. If we abandon methodological naturalism, he suggests, we then open the door to ID theory (which Dilley interprets, for the sake of argument, as a "God hypothesis"),[36] or at the very least, the weakness of the evidence for evolution is exposed. I believe he is wrong about the first point, but raises an interesting point with regard to the second. Let us explore these arguments further.

Noting that it is not a hard and fast principle in science, but one based on tradition, convention, and practicality, ID theorists are very critical of methodological naturalism because it rules out ID as a scientific theory. It does not do so on principled grounds, they claim, but merely for reasons of not wanting to go against scientific tradition (at least on the surface; the real reason, ID theorists often suggest, is that many don't think ID theory is true, but that, of course, is not a good reason to pronounce that it is "unscientific"). ID theorists insist that, when studying physical reality, it is wrong to rule out inferences to the presence of design in nature (and therefore to some form of intelligence) *if the evidence warrants this conclusion*. It is an unnecessary restriction on science, they insist, if we observe evidence of design in nature but ignore it because we are committed to methodological naturalism. One might deny, of course, that there is evidence of design in nature, but that objection misses the point because methodological naturalism seems to imply that evidence of design, even if we were to discover it, must be ignored when doing science (and so seemingly rules out design by definition). It comes close, it

to "dysteleological evolution" (38–42) but discusses it only in the context of atheistic naturalism, and not in terms of how a theist would accommodate the chance elements that are claimed for the operation of evolution within a theory of divine purpose.

36. Dilley, "How to Lose a Battleship," 616.

seems, to seeing what you wish to see in nature, or only looking for what you wish to find, thereby compromising the search for truth.

Dilley argues that many scientists and indeed theistic evolutionists who defend methodological naturalism wander into two big mistakes. The first is that they make *theological* claims (often without realizing that they are doing so). Second, they frequently treat ID and creationist accounts as *scientific* theories, despite their official rejection of this categorization. Both of these moves would seem to be prohibited by methodological naturalism, and so critics are violating a convention they insist everyone else should follow! Dilley thinks that theistic evolutionists who wish to make theological claims when doing science or who wish to evaluate ID theory or creationism as scientific theories must therefore set aside methodological naturalism. Alternatively, they can retain methodological naturalism and refrain from many of their usual critiques of ID theory, but in that case, according to Dilley, their defense of evolution and critique of ID theory is significantly weakened.

Although his essay is boldly subtitled, "Why Methodological Naturalism Sinks Theistic Evolution," Dilley's claim on its face could not be correct, if we read it strictly to mean that if we adopt methodological naturalism, then theistic evolution could not be true! Surely, the way we define and practice science has little to do with whether God brought about life and creation by means of an evolutionary process? Dilley defines theistic evolution as the view that God planned or guided the evolutionary process. This is correct, but it needs a bit more nuance since many theistic evolutionists hold that, although God directs evolution, the actual process also operates with a large element of chance (the view of Collins). Dilley goes on to suggest that most versions of theistic evolution also claim that "God's design of biological phenomena (or history) is *not* empirically detectable using the rigorous methods of science."[37] This may be true of most versions, but again Dilley needs to state this point more carefully. If one follows the convention of methodological naturalism, then Dilley is correct, but this must be understood to mean only that if one examines the human eye, for example, and notes that it appears to be designed, one would not regard this conclusion as *scientific*. One would, however, regard it as both empirically detectable and as a reasonable conclusion, just not one that would be classified as "scientific." He is correct, however, that most theistic evolutionists (such as Francisco Ayala, whose

37. Dilley, "How to Lose a Battleship," 598.

work we will consider in the next chapter) hold that when we examine the eye (or other biological phenomena), the evidence shows that it came about by means of a random process, just as evolution claims, with no evidence of design. However, Ayala, Collins, and others do believe that design is detectable in some way in nature *overall*, and of course, this design must be detected empirically (they are just a bit vague about what kind of design they believe is rationally detectable, but rule out the conclusion that *evolution* is designed).

As I noted earlier, I will argue that design is detectable in evolution and in all of nature because *nature follows a deterministic path.* So, my view is that design *is* empirically detectable, and since I accept the convention of methodological naturalism, I would not insist on classifying this conclusion as part of science. I do insist that it is a *reasonable* conclusion. Collins and other theistic evolutionists, including Lamoureux, at least agree that design is detectable to some extent. Let us not forget that most versions of the traditional argument from design are based on the premise that design is empirically detectable in nature (e.g., in the laws of physics) and that the inference then to a designer is reasonable. Which types of design are present in nature, and whether they include evidence from evolution, is an important, but secondary, matter. It seems that one of the blind spots (no pun intended) of many ID theorists is that they do not appreciate, or don't recognize, the distinction between a conclusion that there is design in the universe arrived at as part of a rational inductive argument and a conclusion of design arrived as part of scientific work. (Whether the former should be included in the latter, since it is based on an empirical argument for the most part is, of course, a central issue, but nevertheless it should not obfuscate the important distinction between the rational and the scientific.)

We might illustrate with an analogy that ID theorists are fond of—that of receiving an intelligent message from outer space (very well dramatized in the movie, *Contact* [1997]). If we adopt strict methodological naturalism, we would have to say that the conclusion that the message comes from an intelligence is not a scientific conclusion. (We would likely have to clarify what we mean because we are so used to mistakenly equating the scientific with the rational that to say it is not scientific is often taken to mean that it is not rational!) While it may not properly be part of the domain of science, it *is* a rational, logical conclusion. Yet, science can obviously help us to continue to study the message even though we conclude that it came from an intelligence. It is similar in biology, it seems to me—if we decide that some parts of nature were designed, or that nature overall was designed, we can still study much of it using the scientific method, which is a branch of rational inquiry. Science helps us understand some aspects of reality, but not all of them. *Indeed, this has been the way that science has operated since its inception.*

Nevertheless, Dilley contributes a very interesting observation, but despite it, he is incorrect that its practical effect is that methodological naturalism seems to sink theistic evolution. One of his chief arguments is that many who critique both the general claim that species and their properties seem to be designed (e.g., the eye) and the more specific ID claim that some direct intervention from an intelligent designer better explains the patterns we see in structures (such as DNA, the eye, etc.) *appeal to claims about the nature of God and his intentions in creating the world* to support their view. This approach, Dilley rightly points out, can even be found in some biology textbooks, as well as in Darwin himself, and in recent thinkers such as those we have already mentioned, Collins, Miller, Ayala, Stephen J. Gould, along with a host of others.[38] Collins, for instance, uses a theological argument to critique ID, saying that God would have designed a "completely ideal" eye, and uses this claim to argue that evolution operating serendipitously better explains its present structure. Dilley retorts that, according to methodological naturalism, Collins should not argue this way—he should not be introducing an argument about *what God would or would not have done* into a scientific debate (his view also seems to indicate that we are unable to show scientifically how the eye came about; otherwise, he shouldn't need to appeal to how God *would likely create*). Collins also ignores the doctrine of the fall, according to Dilley, a well-respected theological argument that is available, meaning that he is implicitly assuming that God would not create nature with flaws and imperfections (some of which we see in evolution). Moreover, he does not clearly explain how it is that nature, with all its flaws, is nevertheless designed overall (which is his view). So Collins' view, as Dilley notes, clearly violates methodological naturalism.[39]

38. See Dilley, "How to Lose a Battleship," 606–8.

39. Further examples come from Miller, who treats Behe's "irreducible complexity" hypothesis as a *scientific* hypothesis, and Alexander, who defends methodological naturalism using the *theological* argument that "it protects God's ontological preeminence and explanatory uniqueness," ensuring that he is not treated simply as another explanatory element within creation; see Alexander, *Creation or Evolution*, 216–18; also, Dilley, "How to Lose a Battleship," 624.

We will come back to the general problem of flaws and evil in nature in chapter 7, but the important point, according to Dilley, is that as scientists committed to methodological naturalism, these thinkers are not allowed to base any part of their argument for their claim that structures such as the human eye arose by chance on claims about God's nature or intentions, on how God is *likely* to create an eye or anything else. God's plans and intentions are very difficult matters to make a judgment about (and any account of them would also require developing a general theodicy). Moreover, the theological judgment of their ID opponents is just as worthy as that of scientists like Miller and Collins, who are not theologians, but this move is ruled out by methodological naturalism. If ID theorists cannot appeal to a designer to explain the design, nor can scientists appeal to what a designer (especially an omnipotent one) would or would not do in order to rule out design.

Dilley suggests more controversially that if thinkers like Collins, Ayala, Gould, Alexander, and others dropped their theological claims, then their empirical claims would not follow. I don't think he is correct about this. It seems that evolution understood as supporting an atheistic conclusion (or as occurring with no design or purpose overall) would not follow, but evolution understood as operating locally without any elements of design or purpose could still be true. In short, the main scientific claim that natural selection occurs (operating randomly) or that common ancestry is a reality—these might still be true, but the further inference that atheism is true would not follow. Their *empirical* claim is not that God would not do things in a certain way, but that natural selection occurs and is operating by chance (and for some thinkers, that this supports atheism). We have, however, noted that their case that natural selection is truly operating by chance is weaker if they do not appeal to the theological claim. Collins might revise his argument somewhat so that he is not saying that "God would not design an eye that way" (i.e., he is not offering a theological argument) but rather is saying only that "the eye has a flaw," and this suggests it is not designed. However, this is a weaker argument, as William Paley argued—since we may still reasonably judge that an object with "flaws" is nevertheless designed if the evidence warrants that conclusion.[40]

It is certainly a very interesting point Dilley raises—that if you insist when studying nature, including biological phenomena and the evolution

40. See Paley, *Natural Theology*, Part 1, ch. 1.

of life, that we may entertain only naturalistic explanations, then you are committed in essence to ruling out, to put it negatively (seemingly by definition), any conclusion that there is design in nature. This then amounts, according to Dilley, to saying that any logical inference you make about causes must conclude that they are *natural* occurrences, which means *not designed*. So to illustrate, say you were studying the human eye, and it crossed your mind (like it did Paley) that it appears designed, but you were committed to methodological naturalism, what would you then have to conclude about its development? It is not clear that Dilley's view follows. It seems to me that you would have two choices: (a) that even though the eye looks designed, it cannot be, and we must seek a natural explanation; or (b) that structural evidence indicates that the eye is designed, and so it probably is, and since science only deals with natural explanations, this part of the structure of the eye cannot be explained by science.

The first approach is ruled out on logical grounds because it is a statement of metaphysical naturalism, not methodological naturalism. It is fine to look further for a natural explanation, of course, and not to jump to a design conclusion too quickly, but if we have satisfied this condition, then to rule out by definition a design explanation overall is to adopt a position of metaphysical naturalism. Even though one might be convinced that naturalism of the metaphysical variety is true (and it might be one's official approach to reality), one cannot call on it here to rule out a design inference by definition since this would be to beg the very question at issue. Nor can one dictate that science commit to metaphysical naturalism, since the question of whether or not metaphysical naturalism is true is disputed and an open question.

The second conclusion simply adds the eye (and perhaps by extension many biological phenomena, including DNA) to the list of things that science cannot explain. This is not a new move or a groundbreaking claim, since there are already several vital features of the universe on this list, including the question of its ultimate origin, the origin of life, the laws of physics, consciousness, and other features of human life, including free will, moral agency, and the existence of objective morality. Indeed, Dilley makes a telling point when he argues that what he calls bench sciences are more successful at giving naturalistic explanations for phenomena, yet when we extend science out to more theoretical work, dealing with and invoking dramatic theories (in cosmology, origin of life studies, evolution, and the theories relating to the nature of consciousness), our conclusions are far less certain, indeed much debated.[41] This work also strays closest to metaphysical naturalism since researchers often seem to want to rule out non-scientific explanations by definition (or they wish to pass off a bias for metaphysical naturalism as a scientific conclusion). It is not a good objection to this claim to reply that the theory of evolution shows us that the eye evolved in a naturalistic way. This is because the evidence to support such a claim is much in dispute. As an evidential dispute, it does not refute the larger point that there are some issues that science cannot explain; it only disputes how many issues of this sort there are. Nor is this a dispute about the nature of science itself, since those who claim the eye, for instance, does not have a natural explanation may still generally accept methodological naturalism. Nor does it settle the more general debate about whether metaphysical naturalism is true since whether the eye or biological phenomena in general have a natural explanation does not settle the question as to whether *everything that exists has*. It is only a dispute about this particular phenomenon. So, if a thinker insists that evolutionary explanations rule out design in the eye, this is a limited criticism, if true, and does not address or settle the larger questions.

Supporters of the claim that evolution operates largely by chance are supposed to demonstrate this claim by making a convincing case that biological phenomena, such as the eye, did in fact come about by chance after all. But the weakness of their case then becomes very obvious, one of Dilley's main points. In the first place, on the general question of whether chance operates at all in biology and the universe, they are often silent and have not considered the question closely, but simply assume that it does. As I have noted more than once, this controversial assumption is one I reject, and it carries enormous implications for this topic (the subject of our next two chapters). But the second, more specific point is that it has proved very hard for evolutionary theorists to show how various biological developments came about by chance, as we noted above. They are often reduced to the "just so" stories or fictional cases. Illustrating by a "story" how evolution *might* have occurred is not evidence that it did occur in a certain way; fictional cases are not a substitute for real cases, particularly in such an important dispute, one relating to the very essence/kernel of the argument. The latter would support a scientific account; the former must be seen as speculation (however plausible it seems).

41. See Dilley, "How to Lose a Battleship," 628.

Evolutionary theorists have piecemeal bits of evidence, along with an *assumption* about chance, that they believe gets them to the conclusion that the "designs" we see came about by chance. However, without actual cases, these claims must remain provisional. That is why so many people of good will look at the claim that organisms are designed and the claim that they came about by chance side-by-side, and are more convinced by the former, in the absence of clear demonstrative evidence for the latter. (And I think we must be especially vigilant about looking at nature and seeing "design" in it but ruling it out because of a *prior* judgment that it came about by chance.)

Dilley also notes that, on the other hand, sometimes opponents of ID theory will often treat ID as a scientific theory, and try to evaluate it as such, despite their official view that it is not science! Miller is best known for this approach with his attempts to show that Behe is mistaken in his claim that we cannot explain irreducible complexities by means of natural selection.[42] However, Dilley is wrong to claim that "with methodological naturalism in play, theistic evolutionists cannot draw upon scientific evidence to critique any God hypothesis,"[43] which in effect seems to be saying that scientific evidence could not emerge that would rule out ID theory. It might be that *current* evidence cannot rule it out, but in principle, evidence might emerge that would. This is because it is possible for a reasonable claim, but one that is not part of science, to be ruled out by further evidence that comes from science. For example, if I speculated that a leak in my car tire was caused by my neighbor leaving glass debris in the roadway (a reasonable claim, but not a scientific one), this could be refuted if a scientific study of the tire showed that it was manufactured with the wrong type of rubber in the value stem, a type that deteriorated rapidly, allowing leaks to develop. So Dilley is incorrect to say that if ID theory cannot be assessed in the domain of science, that it is not then subject to scientific refutation (whatever about it being subject to refutation at present, given current evidence).

So overall, although Dilley is right about his two key points—that evolutionary theorists and other opponents of ID often inconsistently rely on claims about the nature of God and his intentions, and also often do regard ID (and by extension creationism) as a scientific theory and then try to refute it—it does not follow that we must abandon theistic

42. See Miller, *Finding Darwin's God*, 130–64.

43. Dilley, "How to Lose a Battleship," 618.

evolution. As long as we refrain from making theological claims and treating ID as an alternative scientific theory, then we can adopt the approach of methodological naturalism. We could still try to refute ID and any other alternative proposed by appealing to the scientific evidence only. And even if we did not regard ID as scientific, we could not rule out that it might still be true, as many have noted, since whether or not it is scientific is a debate only about how it should be *classified*. In a similar way, "the laws of physics" argument from design, or the fine-tuning argument, might not belong to the domain of science, but are still proposed as reasonable arguments with regard to the question of the design of the universe. Dilley, however, is right that many today who are theistic evolutionists would have to change their whole approach given his key insights. And when they do so, the debate about design in evolutionary patterns must be engaged anew, as it were, since the weakness in their view is exposed. Critics of ID theory could, of course, pursue a line of debate at the level of theology (in addition to the domain of science) by arguing that God would not have designed things the way we find them in nature. But this would dilute their views, as we have noted, since they are on weaker ground overall trying to argue the issue theologically. And if they appeal to science as well to support their theological views, the problem then is that the scientific evidence would not offer any stronger foundation. They also cannot argue that other theists are ignoring the science—since most critics are only questioning some of the evidence for evolution (claiming that it does not show the presence of chance, and not claiming, like many of the ID theorists, that it did not occur at all).

So I accept methodological naturalism as a good convention in science, since it largely keeps science and non-science distinct, though, as Dilley notes, the demarcation line is still hard to draw. Dilley is correct that we must acknowledge the limits of science, which will also help us avoid slipping into metaphysical naturalism. It should really go without saying that methodological naturalism does not commit one to metaphysical naturalism, nor does it rule out design by definition, as I have already noted. To commit to methodological naturalism in science is not to commit to the view that nature operates by chance. This is where Meyer and Dilley overstep the mark with their claim that one is essentially committed to this view if one is a theistic evolutionist. In fact, I would suggest that most who are theistic evolutionists (but admittedly not most contemporary theorists) do not think that nature came about by chance, nor do they think that evolution is largely governed by chance (in the strong sense that human beings might not have emerged from the process).

A commitment to methodological naturalism is not simply a convention, however. There are some good reasons for holding it as well. First, it is sensible to seek a naturalistic explanation in most areas where we are concerned with a study of the physical. A second reason appeals to an inductive argument that has influenced many; we have usually found a naturalistic explanation for recalcitrant physical conundrums (excepting the obvious topics noted above). Dilley rejects this reason for being too selective historically,[44] but it is a stronger claim than that—we have good evidence that we will find a naturalistic explanation for most problems (and we also know the general areas where we might not). Whether or not this includes evolution or the structure of DNA (as Meyer argues) may still be in dispute, but this is not enough to abandon the convention, especially if we are mindful of the distinction between metaphysical and methodological naturalism, and maintain that a design inference is reasonable, even if not included in the restricted domain of science. Third, the convention of methodological naturalism rules out frivolous appeals to design in areas where such appeals are not appropriate, whereas abandoning the convention would allow this. It forces us to look for physical explanations where we might not, as McDonald and Tro, among others, have noted; indeed, this is the way science works.[45] If we cannot find a physical explanation and have a "gap," then we can have a discussion about whether a physical explanation is likely to be forthcoming. All of this can be sensibly handled both within and outside of science.

Our conclusion then is that, despite these interesting and sophisticated arguments of the creationists, ID theorists, and their fellow travelers (and given our qualification about their skepticism with regard to the presence of chance in nature), their critiques and rejection of theistic evolution (understood broadly) are not convincing.

44. See Dilley, "How to Lose a Battleship,"627; also his "Philosophical Naturalism and Methodological Naturalism."

45. See McDonald and Tro, "In Defense of Methodological Naturalism"; also Murphy, "Phillip Johnson on Trial," 33.

3

Chance, Evolution, and God's Design

The phenomenon of common ancestry and the phenomenon of chance generate probably the *two main questions* concerning the relationship between evolution and religion, the two issues that exercise protagonists on both sides the most, the ones that have provoked considerable argument and controversy. But while common ancestry has been the subject of extensive debate, it is quite remarkable how few thorough analyses there are of the crucial topic of chance. Most discussions are superficial, often confused, and frequently misleading. Although scholars will introduce the notion into their thinking about evolution and religion, many give it insufficient attention, and indeed some barely discuss the topic at all. Often it does not even appear in the indexes of books devoted to a study of evolution and religion. This is quite the oversight, given that for most Christian theists the conflict between chance and purpose raised by the theory of evolution is regarded as a more serious problem than any apparent conflict between evolution and the Bible.

This oversight is evident in any number of works on evolution and religion. It is not so much that thinkers fail to raise the topic, but that they address it insufficiently, settling for superficial and sometimes glib pronouncements that fail to come to terms with the serious questions raised by the apparent conflict between the lack of purpose in the process of evolution and God's design of the universe and of life. The work of Francisco Ayala (whose views we will consider later in this chapter) is a good example, although he is by no means the only thinker who fails to face the topic squarely. His book on the subject of reconciling evolution and religion, where one would expect the concept to be analyzed in detail, barely mentions it, despite the fact that Ayala's claims about evolution operating by chance are an essential part of his general position, not only on evolution but also on its relationship with religion. The term does not even appear in the index of his book (nor does the related notion of randomness). Another example of the neglect of the crucial concepts of chance and randomness is evident in the defense of theistic evolution offered in the work of Howard van Till. We already noted in the previous chapter the same lack in the work of Denis Lamoureux. Collins, whose views on theistic evolution have garnered a lot of attention at a more popular level, is yet another thinker who glosses over the question of chance. The simple fact is that for many thinkers the existence of chance, its implications for evolution, and its role in the debate surrounding evolution and religion are accepted mostly on superficial grounds. This is particularly the case with regard to its metaphysical or ontological status in the universe in general. It functions for many thinkers (as noted in chapter 1) as *an assumption of evolution*, without consideration of what it would mean if chance were really part of the universe, if it were a phenomenon we needed to take account of in all of our scientific work, how it would affect our results and predictions. It also functions as an important part of many theological views that attempt to reconcile chance with God's purpose in the universe, leading them into proposing what I regard as non-viable positions, as we will see as our discussion unfolds.

With regard to the *metaphysical* question of chance, it is interesting initially to give some reflection to whether anything happens (or indeed can happen) in our universe by *chance*, and, especially, to what we *mean* by this question. If, for instance, an asteroid crashes into the earth, or a boulder falls down into a valley and kills a bear, or various causal pathways come together leading to the formation of the Grand Canyon, or a man taking shelter under a tree is struck by lightning, and so forth, is it true that events such as these can be said to happen *by chance*? We enjoy thinking about a similar question concerning "games of chance," such as scooping the jackpot on the slot machines or swooping in and winning the lottery by chance! If your numbers come up in the lottery (these days selected by a computer), is this a chance event in the sense that the computer could just as likely have picked a different set of numbers? What if a human being reaches into a hat and picks out your winning number? This might seem like genuine chance because it involves free will, but this case is best described as luck rather than chance, since there is an incomplete (rather than a full) scientific or causal process involved due to the role of human free will in the sequence. So while it is a matter of luck that *your* ticket was pulled out, the event still involves an intentional move on the part of the person selecting the ticket. If a machine selects the ticket, I contend that it is not a case of chance either because the process is determined from start to finish by cause and effect, and again, it is luck that it is *your* ticket, but not that *that* particular ticket was selected. Nevertheless, the answer to the question about chance in all of these cases is not obvious, forcing us to look beneath the surface, to examine this fascinating concept in more detail.

Five Meanings of Chance

Let us consider the various meanings of chance in a more formal way, keeping in mind influential historical views as well. Not only will this help us to identify the different ways the term is used, but it will also allow us to hone in on the essence of the concept, and to focus on the most significant understanding of the term for our work here.[1]

1. **The common meaning** of the term is to describe our lack of knowledge of the causes of an event. So when we say that it was "by chance" that a crop was ravaged by a disease or that a species was wiped out by a natural disaster, we mean that we do not *know* the exact causes for these specific happenings. Darwin used the term this way, expressing the view more than once that, "I have hitherto sometimes spoken as if the variations—so common and multiform with organic beings under domestication, and in a lesser degree with those under nature—were due to chance. This, of course, is a wholly incorrect expression, but it serves to acknowledge plainly our ignorance of the cause of each particular variation."[2] Bertrand

1. I have been influenced in this classification of the meanings of chance by Alexander Maar's helpful work on this topic; see Maar, "What's a Chance Event?" For other discussions of the concept, see Ruelle, *Chance and Chaos*; Humphreys, *Chances of Explanation*; McCall, *Macroevolution, Contingency, and Divine Activity*.

2. Darwin, *Origin of Species*, 173; see also Hodge, "Chance and Chances"; Hodge notes that "Darwin held the most common view of his day: that chanciness in causal

Russell also seemed to think of chance in this way, but most of his examples involve human free will, which complicates the matter because these cases do not involve pure physical causation (a point we will come back to).[3] We find also the following remark in Hume, who seemed to hold varying views on causation: "Though there be no such thing as chance in the world; our ignorance of the real cause of any event has the same influence on the understanding."[4] This is another example of the concept being understood in this way. This use of the term, however, is not asserting that *real* chance exists. It only describes our *ignorance* of causes (causes that may not be due to real chance). Ignorance is the main idea here, not chance itself (ignorance should also not be confused with unpredictability, which we will come to below).

2. **A second meaning** describes the intersection of otherwise independent chains of causes in the universe as a "chance occurrence." For example, in cases where an asteroid smashes into a planet, or a river overflows its banks and destroys a bridge. We find this use of the term in the work of theologian Arthur Peacocke (whose view we will consider in chapter 5), and, earlier, in John Stuart Mill, who observed that: "It is, however, certain that whatever happens is the result of some law; it is an effect of causes, and could have been predicted. . . . It is incorrect, then, to say that any phenomenon is produced by chance; but we may say that two or more phenomena are conjoined by chance, that they coexist or succeed one another only by chance."[5] This understanding seems to be what many people mean when they affirm that there is chance in the universe. Yet, it is not clear, as we will see, whether cases like these amount to real chance. *It all depends on whether the independent causal chains interact out of chance or out of necessity*, a question to which the answer is far from obvious. If independent causal chains interact by chance, then this would count as real chance in nature.

3. **A third meaning** is suggested by the second meaning and perhaps is a more thorough version of it. It is chance understood as *the denial of*

---

theories is due not to gappy indeterminacy in the causal order of nature but rather to gappy incompleteness in our knowledge of that order" (40).

3. See Russell, "Reply to Criticisms."

4. Hume, *Enquiry Concerning Human Understanding*, Part II, VI, i.

5. Mill, *System of Logic*, 526.

necessity in the laws of nature. Necessity refers to the idea that there is a *necessary connection* between causes and their effects, which is revealed as we discover and codify the laws of science. That is to say, when a cause occurs under certain conditions, its effect *must* also occur; it is necessary, not contingent. Philosophers, as Hume noted, have found it quite difficult to pinpoint in what this necessity consists, but most agree that, all other things being equal, its presence between cause and effect means that the effect will occur whenever the cause occurs. Acceptance of a role for chance does not usually mean the complete denial of necessity in nature, only its denial in some or occasional cases, so that in cases where chance is said to be present, an effect need not follow a particular cause. This kind of chance is what I refer to as *real chance*; it seems to be the kind of chance that many think occurs in nature, and especially in the process of evolution. We will elaborate on this crucial understanding of chance in what follows. Hume famously questioned the existence of this necessity between cause and effect, wondering whether it was really present in reality or merely an imposition of the human mind on reality resulting from our tendency to arrange and order things based on our study of individual instances. However, our general assumption when doing science is that this necessity is in nature itself; if it occasionally does not hold true, this would be a case of real chance. John Dudley suggests that this type of denial of necessity can also be found in the work of the pre-Socratic philosopher, Democritus.[6]

4. **A fourth sense** refers to the way in which events involve or affect free human beings. Sometimes we talk about chance events that have an impact on us. We need to distinguish this use of the term from the earlier uses in order to make some important clarifications and clear up some possible areas of confusion. We should distinguish between cases involving humans and cases involving only natural occurrences outside of human activity. The reason the distinction is important is because human beings have free will, which plays a role in human action.[7] So our initial approach might be that a person's free choices could introduce an element of chance into the world

6. See Dudley, *Aristotle's Concept of Chance*, 9–16.

7. It is clear that I am assuming a libertarian account of free will in this book. A defense of this view would take us too far off course, but see Swinburne, *Evolution of the Soul*; Kane, *Significance of Free Will*, among other recent defenses.

in the sense that a person freely decides to play the slots and just happens to win the jackpot. If he had not gone into the casino that day to look for his brother, he would never have won! However, this does not count as a chance event, because the slot machine is not operating by chance (as we will see), and *our choice* to play it is not a chance event either. It is a freely made, conscious decision. If, after a man has just crossed a bridge, the bridge suddenly collapses, we would naturally refer to this happening by saying he missed death by a lucky chance! Yet, neither is this a case of real chance, since his *decision* to walk over the bridge was a free decision, and the collapse of the bridge has a scientific cause (and so there is no chance involved). It is a stroke of luck, rather than a case of real chance. It is fine to describe these events as chance events as long as we do not confuse their occurrence with the notion of real chance described above. There is also perhaps a psychological sense of chance that we could identify here, one that is similar but that need not involve any input from free will. This sense refers to events that occur suddenly, unusually, or unexpectedly in human experience, whether they involve free human decisions or not. These events might include cases where a boulder falls on a dam and destroys it but just misses your car, or where a gas explosion destroys a line of buildings but not the one in which the town residents are holding a meeting, and so forth. We can correctly and usefully describe such occurrences in the world as "chance events"—observing that "it was pure chance that their building was not destroyed"—as long as we do not mean to conclude that there is no complete scientific explanation for why the gas explosion did not reach a particular building. These are not cases of the absence of necessity described above since the causes involved in these events are fully deterministic (the composition of the gas explosion causes it to have a certain range, and so forth).

In this respect, we might consider Aristotle's discussion of chance, since he offers the occurrence of unusual events as examples.[8] He also introduces other cases for us to consider. Most of his cases don't seem to distinguish clearly between events in the world of inanimate objects and in the world of human agency (or their interaction). He describes the case of a man going to the marketplace and just happening to run into another man who owes him a

8. See Aristotle, *Physics*, Books II and III.

debt as a chance meeting, since neither went to the market with this intention. Yet, while this happening may be described as a chance event, it is not an example of chance occurring purely in nature, since both actions here are under the influence of human free will and are not causal events in the world of inanimate objects, subject to scientific law. The existence of this kind of human chance does not show us that real chance occurs *in nature*, nor does it commit us to its existence. Sometimes Aristotle describes this type of chance as "luck" or "fortune." His view has been the subject of much debate, and it is not clear if he thinks that chance exists as something real in nature; part of the difficulty is that most of his examples involve human agency, which complicates the issue.

Aristotle's reflections turn our focus to the notion of intention and its relation to the idea of chance. With regard to human agency, it is easy to see the connection since we can only describe the meeting in the marketplace as a chance meeting if neither party intends for it to happen. But how would this notion apply to the world of nature outside human agency? We need to be careful here because it would not be correct to describe a chance event in nature, say a mutation that affects the path of evolution, as one where no intention is present. This is because the event in question might still be caused in the scientific sense of necessity, whether or not it was intended, so it would not then occur by chance. And if it was the type of event that man or even God intended to bring about, then the action that was prompted by the intention would be a causal event that starts off a causal chain and so not an example of chance. We cannot, in short, define chance as the absence of intention, except perhaps when referring to the human meaning of chance, or as it applies to human actions. Aristotle is concerned with the place that chance would occupy in his overall metaphysical scheme, and, in particular, how it relates to his account of the four causes of things.[9] He concludes in the end that chance itself cannot be understood as a different type of cause, as a power or a force in nature, since this seems contradictory; he understands chance mainly in relation to his four causes; in particular, that it is opposed to the notion of final cause. But his

9. There is considerable debate about how to understand Aristotle's views on chance and how they relate to (metaphysical) determinism, since he approaches the notion from the point of view of his four causes. See Dudley, *Aristotle's Concept of Chance*, 271–324, for a full discussion; also, Aristotle, *Physics*, Book II.

conclusion does illuminate the difficulty we face with the notion of chance. Real chance would be a kind of "power" in nature, since it is purported to be a real part of nature. We might also think of it as a kind of "scientific chance"—an oxymoronic term to indicate that it is real, seems to lead to effects, and is independent of human agency, yet it cannot be scientific in the usual sense since it is understood as the absence of (or outside of) necessity in that some prior causes (that are subject to chance occurrences) do not then lead to certain (and often predictable) effects (some effects can occur but not from their seemingly prior causes, and so forth). *This is why it is difficult to make sense out of the notion of real chance, and part of the reason I will suggest that there is no chance of this type in the universe.*

5. **A fifth sense of chance** refers not to chance itself but to *unpredictability*. It is essential to distinguish between unpredictability and chance, and to consider unpredictability as a category in its own right. It seems that many thinkers confuse or conflate unpredictability with chance, and often use the terms synonymously. This confusion is rife in the literature on these topics; the concepts are often run together when they clearly should be kept separate. However, logically speaking, it is clear that the various steps in what is a fully deterministic causal process might not be obvious to us, and so might be unpredictable in advance. Yet it should be clear that *it does not follow from our inability to predict an event that its occurrence is not determined completely by prior events*, that there must be an element of chance in its occurrence. I think this distinction is familiar and acknowledged by most, yet it is often overlooked in discussions of chance and evolution.

   Sometimes the distinction is described in more technical language as metaphysical or ontological chance vs. epistemic chance, where the latter is simply a failure at the level of human knowledge, but the former describes a real indeterminism in nature itself. The reason it is hard to predict events in advance at the level of human knowledge is because there are often too many variables involved that are difficult to measure. Predicting, for instance, exactly when an avalanche might occur, threatening a dam, is extremely challenging, perhaps impossible. *In principle* it can be done, but the problem is that *in actuality* the variables involved in the calculations are not fully discoverable in advance. We might need to know, for instance, just how much force is needed (from water, wind, rain, etc.) to move rocks and precipitate an avalanche; we can also calculate in principle the damage this force can cause at a certain speed on an object in its path composed of a specific material, weight, etc. (e.g., the dam), but we are unable to calculate these variables with precision in complex causal chains. We can make only approximate predictions; for example, that the conditions are present for a likely avalanche in the near future that would severely damage and perhaps compromise the dam. Much of science works this way. It assumes deterministic chains of causes, but often cannot predict with certainty whether and when an event will occur (though of course in many cases we can, such as in the operation of machines, and even, increasingly now, when we can expect rain!). As a simple matter of logic, there is a clear difference between the occurrence of an event in the world and whether or not we can predict or know that the event is going to occur. We will examine this distinction again later and look at how scientific advancement is trying to narrow the gap between deterministic outcomes and unpredictability in many areas of life, including meteorology, medicine, diet, environmental changes, communications, space travel, and so forth.

Some thinkers, such as C. S. Peirce, have interpreted this lack of precision to reflect an indeterminism in nature itself, but it seems clear that this does not follow. Peirce understood chance as a kind of objective probability. In scientific research, we are restricted, he suggested, to an approximate calculation of the probabilities of an event occurring, so chance then would be the indeterminate element that we are unable to specify and that would make a difference to whether or not an event actually occurred. However, surely it does not follow that this indeterminate element *is in nature itself*; it might be just a further instance of our ignorance of the determinate causes in nature, chance understood as ignorance (or "epistemic chance"). It seems that Peirce's move to indeterminism is not justified by the evidence from science, and surely would not follow just from failure to *calculate* the probabilities of events in nature. The increasing precision and reliability that have been achieved in scientific practice are arguments against Peirce's view, despite his general focus on what happens in everyday science. He is reduced to describing scientific laws as "habits" (which sounds as if he means psychological projections) in an acknowledgment of the

> obvious law-like behavior of nature, though his general point that exact precision is never possible (whether it is due to indeterminism or our always incomplete knowledge at the epistemic level) is a sober reminder of an ultimate limit to scientific prediction.[10] Some thinkers point to events where only probabilistic accounts can be given, even in principle, such as those involving radioactive decay and photon emission and absorption in quantum mechanics, and suggest that this shows an indeterminacy in nature. This brings up the question raised by Walter Heisenberg's uncertainty principle in quantum mechanics—whether the indeterminism is in nature itself or just reflects an inadequacy in our methods of measurement, even if it is an inadequacy that we cannot fully overcome. This question, however, remains open. We will return to this fascinating topic in the next chapter when we consider an objection from the nature of quantum mechanics to our view that nature is deterministic.

So we see that it is vital not only to distinguish these different meanings of chance, but also to emphasize the key distinction between *real chance* and *apparent chance*. Cases of *apparent chance* refer to our ignorance of causes and their effects and the necessity between them, or to our inability to predict causal outcomes in practice even though the outcomes follow of necessity from the causes. *Real chance* refers to an absence of necessity in nature in some cases and in some form, to situations where the effect does not follow of necessity from its prior cause (or more accurately, perhaps, where the prior cause could have been different than it actually was, even given its prior cause, a scenario we will analyze below in detail). We must also keep distinct the role of free will when talking about chance; free actions do not count as physical causes in the usual sense, as we noted, and it is also not appropriate or convincing to speak of chance as a good way of explaining the operation of free will. This is because free decisions need to include intention and deliberate human control, and real chance seems to preclude both of these. Free will would be outside the laws of physics as we know them, and should be kept distinct from scientific accounts based on appeal to cause and effect.[11] So

10. See Peirce, "Doctrine of Necessity Examined."

11. Some have suggested that chance would be a way of explaining our free will, but this view is extremely hard to justify for the simple reason that free will, although understood as having independence from physical causation, is nevertheless deliberative and directed by reason, and so aimed at an end in most cases. Any account of free will, therefore, that appeals to chance or indeterminism would utterly fail to explain the

given these distinctions, let us now turn to the way in which real chance is supposed to enter into the process of evolution.

What Chance Means in the Theory of Evolution

When evolutionary theorists and theistic evolutionists and others claim that evolution operates by chance, they mean that there is *real chance* in the universe. This (usually) means, first, that there is no one (such as God) guiding the process of evolution, and there is no one who planned out the process in advance. And so (for *most* versions of theistic evolution) there is no one who knows the final outcomes in terms of which species will emerge, and, by extension, no one who knows the future destiny of life in the universe. It also (usually) means that when a causal chain is initiated in nature, in most cases, there is no predetermined way it must go, no particular pathway that is laid out for it to follow, no special outcome it is intended to bring about. (There are also some views that are in between, that suggest you can allow for some kind of intention behind the outcomes and yet leave room for an element of chance. We will discuss these and related views in later chapters.) *So this understanding of evolution is committed to the position that there is real chance in the universe.* According to current versions of evolutionary theory, the outcomes of the process are *not* preordained by the initial ingredients of the universe together with the scientific laws, despite the fact that such a view is very controversial and is not generally followed at the practical level of doing science in the domain of physics. If this view were also adopted in physics and not just in biology, it would mean that ordinary natural events in the universe, such as an asteroid hitting the earth, or a storm wrecking havoc on the ecosystems of a forest, or an avalanche destroying a dam, and so forth, *would be subject to an element of chance. And so these events might not have happened.* If real chance of this sort does exist in the universe, it would mean that evolution does not follow any intended path, that no outcome is planned in advance or must come about.

Sometimes this point is made in biology by arguing that when a mutation occurs in DNA, or an environmental change that has an effect on an organism or a species, these events occur *without regard to the well-being*

---

way that phenomenon works in our experience, especially since a "free decision" that occurred as a result of chance would leave us with no control over the decision (the same problem that confronts God's action in a world governed by chance, in fact). We will return to this topic in the next chapter, and again in chapter 7.

of the organism or species. More specifically, no mutation in the DNA of an organism or no environmental change must lead (or is intended to lead) to an advantageous or good outcome for an organism or a species. It might do so, and it might not, since the future pathway and so future outcome is significantly governed by (real) chance. To hold such a view, one must also commit to an indeterministic universe. *Indeterminism* is the view that events in the universe are not predetermined by their prior causes, and can occur randomly or by chance to some extent. This view is the opposite to *determinism*, the thesis that every present state of the physical universe is caused directly (or is completely determined) by initial prior states together with the laws of science. Determinism says that given the laws of science and the initial ingredients in a causal system, the next state of the system has to follow *by necessity*. And in principle, then, the future state would be completely predictable, *all other things remaining equal* (i.e., if there is no outside intervention from God or humans). Indeterminism denies this and holds that the next state does not have to follow, even given the laws and the prior ingredients, because there is an element of chance operating in nature where prior causes *do not always* lead to necessary effects. And so the future is not predictable on this view, even in principle; not just because of our ignorance or inability to know fully all of the variables but because nature is indeterminate.

Yet, most of the scientific evidence is against an indeterministic universe, especially in ordinary physics and biology. (It might be a reasonable thesis to hold regarding causes and effects at the quantum level, though even there this conclusion is controversial, a topic to which we will return.) Why does this interpretation of the way evolution works require indeterminism? Because suggesting that chance plays a role in the evolutionary process requires that a biological event in a biological causal chain might not have happened, even given its prior causes, and this claim requires commitment to an indeterministic universe. So, we can only conclude that a complex organ like the human eye came about significantly by chance if we hold to indeterminism. In a deterministic universe, by contrast, the organ's existence would be foreordained by the set-up of the initial ingredients and the laws of physics by which they operate. (I will argue for a deterministic universe more fully in our next chapter.)

Not only is this claim about chance a very significant claim with regard to the process of evolution, but it is also quite a radical claim in general about how the universe behaves. Chance is such a central notion on both sides of the dispute between religion and evolution due to the fact that its existence appears to be incompatible with the claim that evolution is directed or designed by an intelligence. Atheistic naturalists then use this latter claim to argue that evolution is an argument for atheism because it provides strong evidence that there is no designer involved in the development of nature. Biology as a discipline, without much discussion and mostly as an assumption, also understands the theory this way, and evolution is therefore taught in our schools and universities as a process that is largely governed by chance and that operates without purpose (i.e., there are no *teleological* goals involved in evolution). Yet, it is not so clear that chance is essential to the *true nature of evolution* (rather than current dominant approaches to the theory); moreover, commitment to real chance in evolution stands in tension with broader scientific practice, as we will see.

In response to claims about the role of chance in evolution, creationists and ID theorists *reject* the theory as currently formulated because it relies on chance as the driver of evolutionary change. Atheists insist, on the other hand, that the evidence from a study of evolutionary processes shows clearly the presence of real chance in nature, and that this cannot be ignored in our understanding and exposition of the theory (though in their works they mostly assume it, with only superficial discussion).[12]

Some theists are impressed with the view that chance operates in the universe. They try to embrace the notion of real chance yet show that there may still be room for God and design in the story (as we noted in the previous chapter). Many others, however, are more skeptical with

12. It seems that many textbooks in biology dealing with evolution overstep the mark on how the theory is presented. Given the controversy surrounding the theory and the cultural debates it gives rise to, I suggest five guidelines to highlight in textbooks and in the teaching of the theory: (1) Evolution is a scientific theory not a religious or philosophical worldview, and so we do not make religious or philosophical claims in the exposition of the theory. (2) This book contains no claims about God's existence or non-existence, God's nature, how God is likely to act, or not to act, etc. (3) While evolution is thought by many to have implications for topics in religion, philosophy, and ethics, such implications are not part of scientific inquiry, and this book takes no position on these matters. Scientists have expertise in evolution, not theology or ethics, and they take different positions on the broader questions. (4) Many thinkers describe evolution as occurring by chance (for example, in random mutations). Whether this chance is real or only apparent, or whether evolution is directed toward a purpose, lies beyond what science can resolve. Earlier textbooks sometimes erred by asserting as a fact that evolution operated by chance, and in sometimes making claims that properly belong to theology. (5) The authors of this book believe that the theory of evolution is strongly supported by the scientific evidence (but wish to avoid philosophical and theological overreach!).

regard to the question of whether evolutionary evidence is decisive with regard to the (alleged) presence of chance in the process. These latter thinkers seem to be in general agreement with the atheistic view that the presence of chance in nature, or at least too much chance, cannot be reasonably reconciled with the existence of an overall designer without adding in unrealistic premises or *ad hoc* moves, and so they are then inclined to reject the theory of evolution. So we can see that the question of chance quickly becomes one of the central issues, perhaps *the central issue*, with regard to evolution and religion.

A brief illustration and reminder of how chance is supposed to work in biology will be helpful. A typical approach is to be found in the work of evolutionary biologist Francisco Ayala. Ayala, though committed to the view that evolution operates by chance, is a theist and wishes to reconcile evolution and religion. Although he does not discuss the topic at all, Ayala is an indeterminist about nature; he thinks that even despite the initial ingredients of the universe and the laws that govern them, the evolutionary process could have gone differently than it did in fact go. If this is possible, then there is real chance in nature. So he is working with the standard understanding of chance in evolution.

Ayala first considers eighteenth-century thinker William Paley's famous argument that the human eye exhibits such complex order that it is very reasonable to conclude that it was designed this way.[13] Paley, he notes, makes a number of claims: (1) The human eye is so complex a contrivance that it resembles a watch. (2) Like a watch, it needs several parts all required to fit together precisely for achieving vision. (3) Rays of light, Paley argues, should be refracted by a more convex surface when transmitted through water than when passing out of air into the eye; he then points out that the eye is structured as a convex surface in fish, but not in terrestrial animals, further evidence of design. (4) He then appeals to the phenomenon of dioptric distortion (optical distortions caused by lenses that otherwise correct refractive errors) and remarks that this is cured in the human eye by combining lenses composed of different substances, substances that possess different refracting powers, a discovery that enabled opticians to construct artificial lenses (initially in telescopes), in the same way. (5) The eye has a complex anatomy: a series of transparent lenses, a retina, placed behind the eye "at the precise geometrical distance at which, and at which alone, a distance image could be formed, namely,

13. See Ayala, *Darwin's Gift*, 6–23.

at the concourse of the refracted rays";[14] it possesses also a large nerve communication between the retina and the brain, without which there would be no vision. This complicated structure, both in its individual bits and collectively, shows evidence of design, he concludes.

One can see that Paley's study of the eye is thorough and that his argument is detailed and serious; it is not based on a general, superficial perusal of the remarkable faculty of human vision, a point Ayala acknowledges. So this seems to be a strong argument against the claim that the evolution of the eye occurred by natural selection operating by chance in a series of successive steps, each of which was advantageous to the organism, a view that seems bizarre in the extreme. How do evolutionary theorists like Ayala usually reply to this kind of argument? They make a number of claims in response: (1) Although it does seem as if the eye could not form by natural selection, it did in fact (even though we cannot show how it did so). (2) The eye is not irreducibly complex (it does not need several sophisticated parts working together at the same time in order to get vision). This latter claim has proved much harder to establish since it requires showing what function earlier parts performed before they were utilized for vision. It is impossibly difficult to discover how such parts came about or what function they performed, and so we are forced to speculate about (and even sometimes exaggerate) the evidence we have. Since intelligent design theory came along, this last approach is the one evolutionary biologists favor. They prefer the much-discussed example of the eye because they think they can show at least some of the steps involved in its development. They do not appeal to the human heart, for instance, because it is far harder to illustrate the successive steps in its evolution.

Paley calls the type of design we see in the eye, "relation"; this seems to be similar to Michael Behe's notion of "irreducible complexity." Paley understands this notion to mean what we would now refer to as "the purposeful arrangement of parts" or even "organized complexity." His conclusion is that an entity that shows clear evidence of design must have a designer. And that which is able to design must be a person, a personality; and acts of the mind prove the existence of a mind.[15] He acknowledges that imperfections in organisms and their parts exist, but he thinks these cannot outweigh the evidence of design. Paley is very aware

14. Ayala, *Darwin's Gift*, 18.

15. Ayala, *Darwin's Gift*, 21.

of the problem of evil, of course, but observes that it does not follow that God's attributes are compromised by the evil we find in nature (a familiar argument before Paley).

In the next chapter, we shall have much more to say about the key notion of randomness and how it relates to, and differs from, chance. For now, when discussing randomness as it is used to describe the process of evolution, as exemplified in the work of Ayala, it is helpful to distinguish between two senses of randomness (a concept which Ayala does not discuss in any detail). The first is randomness in the sense of an event happening by chance. The second is randomness, specifically in the evolutionary sense, which means that changes in organisms occur along the way *that are not aimed at being beneficial* to the organism or species. Such changes might or might not benefit an organism, and so, it is claimed, this leads to the conclusion that there is no teleology in evolution.

How is it then that evolution has led to the emergence of highly organized and very complex species over time? Part of Ayala's reply to Paley is that it is purely a matter of chance; such species could just as easily not have emerged, and if we ran the whole process again, we would get a vastly different outcome (as Gould has argued too).[16] This is the sense in which Ayala is using the notions of chance and randomness in his account of evolution.

Ayala develops his reply by arguing that natural selection does progressively favor beneficial traits, and, constrained by an increasingly hospitable environment, the interaction of species with their surroundings gradually leads to the arrival of more complex species. How does this happen? Although offering no detailed discussion of the notion of randomness, or of the topic of determinism/indeterminism in science,[17] his claim is that natural selection multiplies beneficial mutations and eliminates harmful ones, adding, "It is worth pointing out that increased complexity is not a necessary consequence of natural selection, but emerges occasionally."[18] How would this account for the complexity in the eye? Ayala argues that the human eye did not appear suddenly in all its present perfection, nor did it evolve aimed at its current task; "its formation required the appropriate integration of many steps of favorable mutations. . . . Such mutations gradually accumulated."[19] He refers to the

16. See Gould, *Wonderful Life*, 51–52.

17. See Ayala, *Darwin's Gift*, 55–60; also 76–77.

18. Ayala, *Darwin's Gift*, 64.

19. Ayala, *Darwin's Gift*, 65.

well-known (and indeed well-worn) bacteria example to illustrate how natural selection works.[20]

He also turns to the monkey example to further illustrate the process of natural selection.[21] In his version, the monkey sits at the typewriter punching keys randomly; occasionally he strings together a few letters or words that could make sense in a larger script. We then take what he has typed the first time, put it in a new machine, and he again punches keys at random, but this time around, since he has a few meaningful letter strings and words from his first attempt, his second offering moves a little further toward coherence (or organization)! Repeat this process many times and eventually the monkey's random typings may produce something quite meaningful! If we study just the final product it might look designed (or directed, or the result of an intelligence) but in fact it occurred only by chance. Acknowledging that the analogy is very limited, Ayala nevertheless says it is a good illustration of how natural selection works. The idea is that it produces effects that are currently useful to the organism to a limited extent in its present situation, and which might be beneficial later in a more complex set-up involving the same organism (but where the beneficial traits/developments would be performing a different function). (We will leave a critical observation about this argument until the next chapter.)

This example shows, Ayala claims, that the various steps in the process that produce the eye occur by *chance*. Although he does not use the term and has virtually no discussion of the concept in general, he works mainly with assumptions and commitments to a position on what are still contested questions, he means that *real chance* occurs in the evolutionary process. For example, he thinks that cause X, which leads to effect Y in the human eye did not have to occur. Its occurrence is due to chance, and so Y then might *not* have happened if X had not happened. Indeed, Y might not happen even if X happens, because in some cases the effect does not follow from the cause by necessity. Similarly, for the process that led to X, the cause of X did not have to occur if real chance is possible. This means that the evolution of the eye overall occurred by chance, since *many of the steps* that contributed to its evolution *need not have happened* (just as the letters typed by the monkey each time are entirely random). He also thinks it follows that there is no intended outcome, no goal that the process is aimed

20. See Ayala, *Darwin's Gift*, 59–61; I discuss the bacteria example in my *Evolution, Chance, and God*, 43–44.

21. See Ayala, *Darwin's Gift*, 61–63.

at bringing about. The beginning of the process is not aimed at bringing about the organ of the eye; it just so happens to take a pathway that eventually ends up with a seeing eye, yet it need not have done so, since many of the steps occurred by chance, according to Ayala. Ayala has no problem with this account, saying that, "we can readily understand that the accumulation of millions of small, functionally advantageous changes could yield remarkably complex and adaptive organs, such as the eye."[22] This is one reason why, Ayala suggests, the eye has a blind spot. Interpreted as a blemish, the existence of the blind spot shows that the process is haphazard, piecemeal, messy, governed by chance, not designed. Moreover, he and other thinkers often suggest that a designer would not make an eye with a blind spot, a claim that sounds more theological than scientific, and one we will come back to later.[23]

He considers the commonsense objection that it is not plausible to say that an organ as complex as the eye could have evolved piecemeal, an objection often brought up by the proponents of intelligent design theory as well. Since Behe's notion of irreducible complexity, evolutionary biologists have developed a way of responding to the claim that the eye, for instance, needs many parts functioning simultaneously to work, and that it is not plausible that these parts evolved independently of each other, performing different functions in earlier forms of an organism that were later repurposed toward the single task of vision. Many appeal to the work of Miller in this respect, a strong critic of ID theory, as does Ayala, who argues against Behe's view by appealing to Miller's claim that there are various examples of simpler, earlier eyes in evolution.[24] Ayala claims that forty eyes have evolved independently in nature. He acknowledges that there is a problem in our not knowing the function of earlier parts, those parts that were performing another task in the organism before sight was possible—this is the feature that many find implausible in the way natural selection is supposed to operate. It is an especially vexing problem when expanded to include the coming into existence of the whole human body, that each complex organ developed in the same piecemeal way, with

22. Ayala, *Darwin's Gift*, 63.

23. This is his main reply to Paley (rather than providing a detailed account of how the eye came about by means of natural selection, which would be the obvious way to refute him); see Ayala, *Darwin's Gift*, 22–23. As noted, we will come back to this point later, but it is an example of Dilley's point (discussed in the previous chapter) that this type of objection seems primarily theological, not scientific.

24. See Ayala, *Darwin's Gift*, 144–48.

earlier parts performing different functions than they do now, and then all of the separate organs and overall structure somehow further adapting and coming together in an extraordinarily sophisticated, unified organism. This seems to be a new tactical approach adopted by evolutionary theorists since the arrival of ID theory. (Collins also takes this approach, again relying on the work of Miller.)[25] It is claimed that there are earlier cases in which some of the parts are used for other purposes, parts that are later adapted, due to natural selection, to make vision possible. It is hard to judge whether the evidence here is strong enough to justify this conclusion, or whether this is an *ad hoc* move to paper over a real difficulty with the process of natural selection. It is a vital question, and is highlighted by Paley's view that it is the sheer machine-like complexity of the eye that undermines the claim that it came about randomly.

After laying out his view that evolution occurs by chance in the way we have explained, Ayala then turns to theological issues in his attempt to reconcile evolution with religion. We will return to his views on this topic later.

Chance in the Universe: Nine Key Questions

Ayala's examples and illustrations of how modern evolutionary biology is understood and indeed taught raise a number of questions that we must identify and keep in mind throughout our analysis:

1. Although, like most evolutionary biologists, Ayala basically assumes its role and does not argue for it, we can ask a general question about chance that arises from his account of its role in evolution: Is Ayala correct that there is this type of *real* chance in the universe?
2. Might this type of chance be present in biology, but not in physics and chemistry?
3. Does the evidence from evolution itself show that this type of chance occurs, or does the conclusion about chance depend upon prior metaphysical commitments/assumptions concerning determinism/indeterminism, physics, law, and chance?
4. Would the presence/existence of real chance undermine the notion of design in nature and in evolution, and indeed in the universe (as many claim)?

25. See Collins, *Language of God*, 186–93.

5. An extension of the above question: Is it possible to reconcile this kind of real chance with *purpose* in the universe?
6. Are there any other possible forms of chance that might play a role in the way that nature behaves (such as the ones delineated in the early part of this chapter)?
7. If determinism about the universe is true, and indeterminism false, does this rule out the presence of real chance in the universe (and therefore in evolution)?
8. Is there a danger that the evidence from evolution may be packaged in a certain way so as to accentuate the role of chance in the theory (perhaps for ideological reasons, reasons of tradition, conformity, lack of awareness of the deeper metaphysical questions, etc.)?
9. We should also keep in mind for later what is primarily a theological question: Is Ayala correct that God, whom he believes designed the universe, would *not* design the eye or the jawbone in the way that we find it?[26] An extension of this question is: If God is guiding evolution, would he not then be responsible for the present structure of organisms? (And, of course, even if he is not guiding specific evolutionary outcomes, is he not still responsible for their structure and consequences *overall*?)

It will be important to keep these questions in mind as we develop our arguments in what follows. We will respond to some of them in the second half of this chapter and return to the others, including the theological questions, in later chapters.

The claim that real chance is present in biology is controversial, and it is interesting that few scientists ever discuss the deeper issues, but simply assume that this is how biological processes operate. But the difficulty is clear once we consider our general understanding of physics and the contrast between physics and biology. In our ordinary everyday understanding of physics, and especially in the practice of physics, we assume that there is no chance involved. We can see the truth of this when we consider a case of an ordinary causal happening in the universe, such as the occurrence of rain. The rough explanation we are all familiar with from our school science classes is that water vapor, in the form of a gas given off by the earth, rises and cools, eventually forming into liquid water droplets through a process of condensation. These droplets gather

26. See Ayala, *Darwin's Gift*, 154–60.

in clouds, coalescing into each other, becoming heavier, and eventually falling to the ground in the form of rain. *The essential point is that there is no element of chance involved anywhere in this explanation.* We do not say that the first step occurs by chance, or that if the third step had not occurred, we would not get the clouds. We also don't think that *where* it rains is a matter of chance, since the deterministic process that leads to rain also determines its location. This is why the science of meteorology is gradually getting more accurate—because the more we discover about the processes leading to the production of rain, the more we can study them, and use them to predict where rain is forming and when it will fall. One might argue back that rain would occur in the deterministic way I have described if the first step in the causal process occurs—that, yes, given that step, the rest must follow inevitably—but *that step* might not occur. However, this is not correct because the so-called "first step" is simply another causal happening in the universe, which is itself the outcome of a deterministic process, and must occur necessarily, *given its prior cause(s)*. We will come back to this topic in more detail in the next chapter, as we explain and develop the position known as determinism, including the question of the interaction of causal chains, and illustrate with further examples from physics and biology.

The comparison with physics shows that to claim that biology operates with an element of chance is indeed controversial! We seem to be introducing into a biological causal analysis a key element that does not exist in physics and chemistry (or at least that we do not introduce into the practice of these disciplines). This is a problem since the current understanding is that biological processes are simply physical processes in a more sophisticated form, or existing at a higher level of organization (though this debate is far from settled).[27] Whatever the answer to the philosophical issue, biologists do generally approach their subject matter as if it is reducible to physics, or is a higher form of physics and chemistry. For example, when an environmental change (such as radiation) affects the DNA of an organism, this is explained as a process governed by physics, that is, by the physical make-up of the DNA, the physical make-up of the environmental material, together with the operation of the laws of physics. There does not seem to be anything more going on than this, according to biologists, and they are even very reluctant to acknowledge that the whole may be greater, or anything more than, the sum of the parts (so

27. See Murphy, *Beyond Liberalism and Fundamentalism*, 135–44; also Sellars, *Philosophy of Physical Realism*; Ayala and Dobzhansky, *Studies in the Philosophy of Biology*.

that when we understand the parts we will understand the whole). So to then say that suddenly the causal processes operate by chance in biology when they do not do so in physics is, on its face, very problematic. The claim that evolution operates by chance seems to be equivalent to saying that science operates by chance, but this appears to be false, and is surely rejected in our scientific practice. Moreover, evolutionary biologists don't claim that biology and physics operate under different laws, and in fact, prominent thinkers such as Jerry Coyne and Edward Wilson actually go so far as to deny the existence of human free will using the argument that the brain is purely biological and operates, therefore, according to physical and *deterministic* laws.[28]

Moreover, the debate about whether biology itself is governed by laws is an ongoing one. Just because we cannot predict how biological systems will unfold in nature does not mean that this unfolding is not fully obeying scientific law. Yet it seems that biology acts like any other area of science with regard to this matter in that it tries to improve its predictive success by striving to get closer to the lawlike behavior found in nature (such as we see in Gregor Mendel's laws of inheritance, for example).[29] Of course, some areas of nature may remain unpredictable in practice because of the difficulty of the terrain, but this again is not evidence of the operation of chance and the absence of lawful behavior. Biological systems are very complex, making modeling of them very difficult. I interpret all of those "fudging" terms in biology—such as "grand narratives," "historical contingency," "context-dependence," "complex systems," "self-organization," "environmental diversity," "probabilistic outcomes," and so forth—as euphemisms for the fact that we have not yet (and indeed may never) discover the underlying laws by which evolution, and biology, generally operates. At the very least, this fundamental topic requires a full discussion in biology, but I have suggested that it is usually ignored by most scholars or passed over superficially, and a position is simply assumed on it. To make matters worse, it is taught this way to unsuspecting new students, and it is then employed to take a position on the very controversial topic of teleology, usually to rule teleology out of biology, in fact.

28. See Coyne, "You Don't Have Free Will"; see Wilson, *Sociobiology*, 201–20.

29. For more on the fascinating debate surrounding whether there are such things as biological laws, see Beatty, "Evolutionary Contingency Thesis"; Turner and Havstad, "Philosophy of Macroevolution"; Nagel, *Mind and Cosmos*, ch. 3. For the debate concerning determinism and indeterminism in evolution, see Millstein, "How Not to Argue for the Indeterminism of Evolution."

God, Design, and Real Chance

Given the above, we can now identify how the notion of real chance in the universe *relates to God and his design/purpose*. There are four main theistic possibilities (along with two atheistic positions):

1. The universe is created by God and designed to develop *in a deterministic way*, and there is no real chance in its operations (many theists accept this view and hold that the absence of chance supports the position that the universe is designed, and this is an argument for an intelligence). (We are arguing for a nuanced version of this view in this book.)
2. The universe is created by God *in a somewhat indeterministic way* so that real chance is present in its operations; therefore, the final outcomes of the universe are not under the control of, and so not intended, by God.
3. The universe is created, and although some real chance is present, the overall outcomes are nevertheless directed by God in some way toward a certain specific set of ends.
4. The universe is created by God and designed to operate in a deterministic manner. Although there is apparent chance in its operation, there is no real chance.

The two main atheistic naturalist options are:

5. The universe is not created by an intelligence (God); *it operates deterministically*, and there is no real chance present (except at the beginning, when it came into existence); there is no teleology in the universe. (The universe *overall* is a result of chance, however, in the sense that the initial ingredients, laws, etc., came about accidentally.)
6. The universe is not created by an intelligence and operates *indeterministically*, so there is real chance in its operations. (This is the mainstream view among atheistic naturalists—at least with regard to the discipline of biology—and they believe that the presence of real chance supports the claim that the universe is unplanned, and so also contains no teleology.)

We can now consider how theistic views relate specifically to the topic of divine action:

1. The first possibility is that God designs the universe; it operates in a deterministic way, and there is no real chance present (this is my view—defended in detail in the next chapter). This view, however, *allows* that either God or humans (when humans come along) can intervene in the operations of the universe. God is also in full control of the final outcomes of evolution. These outcomes are directed by God and intended by God (until, that is, the appearance of *Homo sapiens*, when the existence of our free will gives us some independence, a limitation that God places on himself). The amount of independence we have with regard to God's intended purposes is a matter of considerable debate among philosophers and theologians, and we will not pursue this matter here.

2. The second possibility is that God designs the universe, and there is real chance built into its operations (i.e., it operates indeterministically); moreover, God, although he can, does *not* intervene in nature after he creates it. In particular, God is not in control of the final outcomes of evolution, a position defended by Ayala; it also appears to be the view of theistic scientists Miller and Collins, and Fr. George Coyne, as well as theologian John Haught (views we will come back to later).

3. A third possibility is expressed by some theologians who, though affirming indeterminism and real chance, propose that God does intervene in nature occasionally but *without* breaking (or "violating") the laws of physics. In this way, God can still control, *to some extent at least*, the final outcomes. So this is a weaker version of the second possibility. This view seems to be held by theologians Peacocke, Peters, Russell, Murphy, and others. The views of these thinkers are sometimes hard to pin down because they vacillate too easily between chance and purpose (as we will see in chapters 5 and 6).

4. Our fourth possibility above, the view that there is apparent but not real chance in the universe, may seem to be just a different way of restating the first option. While this is largely true, it may also describe a distinct option, depending on where one stands on the question of determinism. This view holds that the chance we appear to discern in the process of evolution, and anywhere else in the universe, is apparent and not real chance. This is because God is in fact guiding the process behind the scenes; while it might *seem* to us to be operating in a random way (especially in evolution), causal

happenings are in fact directed by God all along. We find this view expressed in the work of Craig, Plantinga, and many others.[30] It is not always clear if those who seem to hold this view really do subscribe to determinism with regard to how nature operates after God creates it. If they consistently hold to determinism, then it would be similar to the first possibility.

5. Finally, the two atheistic possibilities deny that there is any intelligent design in the universe, and they may or may not believe that there is real chance present in nature.

Chance and Purpose: Possibilities for Divine Action

Let us elaborate further on God's action in the world as it relates to the four theistic options:

1. In the first view, which holds that there is no real chance involved in the universe and that the universe is deterministic, this would mean that God creates and sets up the universe from the beginning, and it then unfolds according to his design plan. Yet, even though the universe operates deterministically, there are different ways to think about how it might unfold. The "unfolding" could take place in one of two ways:

   A. All subsequent happenings could be built into the initial blueprint at the beginning and then left to unfold as initially planned. No further direct intervention by God is required along the way (so this view would be a rejection of a key claim of intelligent design theory). We should also distinguish here between two types of intervention. There is a direct type of intervention in the sense of God performing miracles (such as saving a person from serious injury in a car accident) or in answering our prayers. Christian theists have always held that God acts in this way from time to time. A second type of intervention involves tinkering with the natural processes operating in nature and with the laws of science *as they unfold* in order to further along the divine plan, such as intervening to move evolution in a more favorable direction to bring about a certain

30. See Plantinga, *Where the Conflict Really Lies*, 67; also 272; see Craig, "Response to 'Mere Theistic Evolution,'" 57–58.

species, or to guide the formation of the eye, and so forth. Christians have normally held that God does not intervene directly in nature in this second way (subscribing to the distinction between primary and secondary causes, where secondary causes are the ingredients of the universe and the laws which operate naturally, and God is the primary cause of the overall operation, the first cause who designed the order of secondary causes). (In chapter 5, we will elaborate further on this notion, and on the various ways of understanding divine intervention in creation, including the proposal that God might be able to influence the course of natural and historical events without directly intervening and without breaking scientific law.)

In the view being explained here, which denies the existence of real chance, I allow that the first type of intervention can occur and indeed does occur regularly, but I am denying the second form. (Though in some cases it might be hard to distinguish between them, the general point still holds.) God answering a prayer may sometimes require intervention in the laws of science, but this is not the same as intervening without a human plea being involved—that is the key difference. So if God answers a prayer or prevents a natural occurrence from bringing suffering on humans, this is fully compatible with the denial of chance in the universe. We are saying that God set up a deterministic universe yet can intervene in it occasionally, of course, if he wishes, but he does not do so, except in cases of miracles, responses to prayers, and similar events in this category. The argument for the claim that God does not intervene in the second way noted is based on the empirical evidence of science (a point we will return to below). (This discussion, of course, prompts a question about the problem of evil: Why does God not prevent more evil than he does? This is an important, though not a new, question, one raised especially by the theologians referred to above, and a crucial question to which we will return in our last chapter.)

Staying with this view, there can also be intervention, however, from human beings, when we come along later, because we have free will that enables us to intervene in causal chains that are otherwise predetermined. So, for instance, human beings could intervene in nature to prevent a naturally occurring forest fire from destroying acres of trees, trees that would otherwise be ruined because of the deterministic operation of nature. This, however, is not a case of real chance since free will is motivated by reason, is deliberative and usually aimed at a goal, but it does mean that the blueprint that God has set up unfolds deterministically *until the appearance of Homo sapiens*. So this raises some interesting questions that we will come back to below: Does God know how the future unfolds *after* human beings come on the scene? Is God still in full control of the final outcomes of the universe after we come along? Does our free will prevent God from having total causal influence over us?

B. To re-emphasize the point, the second possibility for the unfolding of the plan for the universe is the intelligent design view, which is that God sets up the initial blueprint, as it were, lets it proceed, but from time to time intervenes directly into his creation to bring about certain effects. For example, according to ID theorists, God intervenes to supply new material in the development of DNA, as we saw Meyer arguing (in the previous chapter). When setting up the initial conditions, God did not include *every ingredient* in the design blueprint; he chooses to intervene from time to time to add new material, and perhaps also to redirect the future course. Most Christians will agree that God in his omnipotence *could* have done it this way, but I, and other opponents of ID theory, hold that the preponderance of evidence from our study of how the natural world operates in science indicates that he did not. Some thinkers, as we hinted above, also offer theological reasons for why God would not do it this way, suggesting that it is not only bad science but an inadequate theology to affirm later interventions by God. Ayala makes this criticism, and is another thinker who suggests that a problem facing the ID approach is that it would seem to make God responsible for evil. This is why he describes evolution operating by chance as a "gift."

2. The second view of chance holds that God is not in control of the final outcomes of evolution and even of the universe more generally. God does not know the species that will develop because he has placed a large element of real chance into the course of nature. It is not just that, although he remains in control, we cannot from our

human vantage point *predict* which species will come into existence, for this would be apparent and not real chance, since God could still be ordaining and intending all the outcomes. No, it is that God has placed real chance into the process, and so he does not know how nature will unfold in the future in terms of cause and effect. He is not directing it, or laying out a deterministic path for it to follow, or intending to bring about specific outcomes such as, for instance, the arrival of *Homo sapiens*, intended to be at the top of the evolutionary tree in terms of our special and unique qualities. If this were God's intention, he would have to direct the process in order to ensure that it came about. This indeterministic set-up would be a limitation that God would place on himself, of course, in the sense that he could have created a universe without real chance, but according to this view, he did not. And so God gave up control over the future, especially concerning which species come into being, and their various properties. As noted above, this perspective is popular among some prominent contemporary theologians.

We should make one qualification about this perspective. Although many balk at the claim that God might not be in control of nature and life, one might object that this view does not present so much of a problem because, after all, God also loses some control by giving human beings free will. So God, it seems, does not fully control human choices and so loses some power over the final outcomes anyway. However, this attempt to assuage the worry about God not being in full control does not work, since it is only true *after Homo sapiens* comes along. The problem from the point of view of a consideration of evolution is that there is no free will in the early history of species, and so the species that emerge, including *Homo Sapiens*, with our distinct makeup (including reason, free will, and moral agency) *did not have to emerge, because of the presence of real chance in the process*. So the point about free will, while quite significant in itself, would not assuage the general worry here.

3. Some thinkers, including Peacocke, suggest that there can be some real chance in the universe, but that, nevertheless, God is still generally in control of the final outcomes. Although this seems impossible from a logical point of view, Peacocke nevertheless tries to defend this interesting position. He suggests that God could allow some real chance to occur, so that he would not know the final outcomes

in localized cases. However, these smaller systems would then fit into a larger plan that would not be governed by an element of real chance, and where God has full control, and so the final outcomes are intended by him. It would be like saying that the formation of an electrical harness necessary for an engine to work could come about by chance, but then the circuit of which the harness is a part is otherwise configured scientifically so that the engine works in the usual deterministic way. One can see the difficulty since it seems that the nature of the harness must in some way be predetermined so that it comes out one way rather than another. It couldn't have any old configuration for then it would not integrate into the larger electrical system, so it is not clear if Peacocke's view can make sense.

It is one thing to say that there is chance involved in a process that still leads to an *intended* result and quite another to say that there is chance involved to the extent that whatever result occurs is *not intended.* It is the latter kind of chance with which we are especially concerned. It is not quite clear that Peacocke is talking about this kind of real chance, as we will see when we come back to his view later on. But Peacocke does seem to be implying that God knows only the various potentialities present in the unfolding of creation, but not the precise outcomes. His work fits into that category mentioned above that suggests there can be a mixture of chance and design in nature. We will consider such views in detail in chapter 5, but this position does seem to be very problematic on its face since the presence of real chance would seem to preclude intended final outcomes, as we have noted. Yet, I think it is correct to say that most theistic evolutionists (the mainstream view) in the scholarly literature—though not in ordinary life—hold that God set evolution in motion but that the process operates by chance and so God has voluntarily given up control of the final outcomes. However, in ordinary life, most religious people (who would also be theistic evolutionists) hold that God is largely directing evolution, and that the species that come about, including the fact that *Homo sapiens* is at the top of the evolutionary tree, are outcomes intended by God.

4. The view advanced by Plantinga and several other thinkers straightforwardly holds that everything that happens in the universe, at least up to the appearance of *Homo sapiens*, is intended by God, including, for instance, all mutations in DNA. However, because

of the complicated nature of causation and the many variables involved, which contribute to our inability to predict in a very large number of cases which mutations will emerge, it might *appear to us* as if these mutations occur by chance. Our distinction between *real chance* and *apparent chance* is helpful here. Apparent chance describes events that appear to us to be occurring by chance, but that are in fact directed by God all along. We would realize this if we knew as much as God knows, or at least if we could identify the full causal chains involved in natural processes. So on Plantinga's view, God intends all of the outcomes of the process of evolution (if it occurred, about which he is skeptical), though it might appear otherwise due to our limited knowledge.

These scenarios are obviously relevant to whether God knows the future, a question also pertinent to the topic of the plausibility of theistic evolution. In traditional theology and philosophy, God is understood to be omniscient (as well as omnipotent and omnibenevolent). This is generally understood to mean that not only does he have knowledge of the past and present, but also of the future. On this conception of God, he would know the future in a deterministic universe, because he would know every future event that was going to occur, given the laws of physics and the initial ingredients. This view is also consistent with the notion of God's providence. Many thinkers also hold that God's knowledge extends to the course of free human actions even though they remain free. Various theories of God and his relationship to time are often invoked to explore this matter, yet thinkers continue to argue about the degree of knowledge God, logically, can have of the future.[31] We will not take a definitive position on this matter here, but leave it as an open question, since it will not have any significant consequences for our arguments in this book, except to note that in our view God has knowledge of the future up to at least the appearance of *Homo sapiens*, and to a very significant extent afterward. (Indeed, the question of determinism and the question of God's knowledge of the future seem to be conceptually distinct, in the sense that it seems possible that God might not know how the future will turn out, but yet nature might still be operating in a deterministic way.)

31. For various perspectives, see Stump, *Aquinas*; Hasker, *God, Time, and Knowledge*; Grant, *Free Will and God's Universal Causality*; McMullin, "Cosmic Purpose and the Contingency of Human Evolution."

With this overview of both the various understandings of the concept of chance and of what they would each mean for God's role in creation, I wish to explain and defend the first view, that there is no chance operating in the universe, including in biology. We will then consider the other views in more detail in the work of various thinkers in the chapters that follow, and contrast those views with my view in building our case for the viability of our version of theistic evolution.

4

A Deterministic Universe

We need to begin by laying out and explaining our view that there is no real chance present in the universe. This will leave us in a better position to distinguish this view from the various other perspectives noted in the previous chapter and, in particular, from the perspective that has come to dominate current understandings of how evolution works. My thesis is that we live in *a deterministic universe*. Although there are some different ways of approaching the topic, determinism, as briefly noted in the previous chapter, is the general thesis that every present state of the universe is caused directly (or is completely determined) by prior states together with the laws of physics. It is a quite straightforward approach to physical reality, and is a topic that philosophers and scientists have discussed since the dawn of time.[1] The deterministic thesis also provokes consideration of related questions: Is real chance possible? What role does human free will play (when it comes along in the history of life) in the unfolding of events? From a theistic point of view, what is the extent of God's knowledge of future events? Theoretical reflection on the concept of determinism in nature can be complicated and sometimes confusing.

1. For further discussion, see Earman, *Primer on Determinism*; also, Koperski, *Divine Action*, ch. 6; Koperski's somewhat labored overview is a good indication that there is no clear-cut or obvious way forward on the question of determinism/indeterminism in either metaphysics, physics, or mathematics. See also Maitzen, *Determinism, Death, and Meaning*, chs. 2 and 3, for a robust defense of determinism, yet one that ignores the question of how we ended up living in a deterministic universe.

So let us introduce the topic with a few simple examples from ordinary scientific explanation as a more familiar way of illustrating how science is understood to operate in a deterministic way before then turning to more philosophical matters.

Determinism and a Universe Without Chance

Suppose the brakes are failing on your car, and you take it to a mechanic. After an inspection, he tells you that the brake line, which carries fluid to the brake calipers, is leaking, and the resulting lack of vacuum pressure is the cause of ineffective braking. The line perished, he further explains, due to the age of the materials, the excessive heat it is subject to in its day-to-day operation, along with other causes. The mechanic adds that the average life of a brake line is ten years, but that it is difficult to predict when or even if a particular line will fail. There are several variables that are challenging, perhaps impossible, to calculate with precision, such as the specific rate of heat the brake line is subjected to in a particular car and with a particular driver, the degree of stress on the braking system, the exact make-up of the hose and other brake components, and so forth. Even if we could identify all of the variables, it would be quite impossible to measure them from a practical point of view to predict with any accuracy when the line might fail. An experienced mechanic might be able to tell that a brake line is likely to fail "soon," but predicting exactly when, even within a range, is not possible.

The important points to emphasize in this everyday example are: (1) Given the heat intensity, the specific make up of the brake line, the exact rate of pressure on the hose, and so forth, *it has to fail at the time it does* (if we could know all of those things with certainty, then we could predict when the hose would fail). (2) When the hose fails, the brakes cannot work properly due to prior causes, such as lack of fluid pressure and so forth. In other words, *this is a deterministic process*. There is no (real) chance involved in it anywhere along the line (pun intended!). (3) It is always open to us to use further knowledge and research to reduce the risk of a brake hose developing a leak—for example, in producing rubber that withstands heat stress more effectively, *given the assumption of determinism*. That last point is essential. If we allowed that real chance could play a role in the process, then our predictions would be for naught! If we allowed, for example, that the heat intensity affecting the hose did not always obey the laws of physics, but was sometimes subject to real chance, then we would not be able to follow out the causal process with any degree of confidence. The fact is that when doing work in physics, we assume the universe is deterministic and don't usually get into the metaphysical questions. This assumption is not unjustified or frivolous or irresponsible, for it is based *on our experience with the way the natural world works.* We have found that the more we investigate a causal process, the more we uncover its deterministic nature and the more we can control it and make more accurate predictions. This, in fact, is how science works at the everyday level.

We can take another everyday example from the world of medicine and healing. We know, even if we don't give it much thought, that drugs are made up of combinations of chemicals. These chemicals have been tested over time to determine how they affect certain body cells, so we can be confident that if we apply an ointment to a rash, there is a high degree of probability that it will clear up the ailment! We know that there are scientific laws for why this is so. Just as important, if the ointment does not work, we know that this is not because the scientific laws did not hold or that the cause-and-effect process did not follow a deterministic path; it is because there was an unknown or unsuspected or hidden variable in this particular case that affected the results. The patient might have been using the ointment too sparingly or inadvertently washing it off too quickly, or the ointment may have been accidentally exposed to a heat source in its production, rendering it less effective, and so forth! This causal determinism in the universe makes the predictability that is becoming more and more accurate in medicine (and many other fields) possible.

However, sometimes we are not quite as tuned in to the fact that the universe is operating according to scientific laws at the practical level of everyday living, even though we accept causal determinism. If, for example, the kitchen sink is not draining, we quickly conclude that it is blocked. Here, we are appealing to causal determinism—to the fact that a solid enclosed mass of a certain density (the blockage in the pipe) can prevent the passage of water, causing a backup. We infer that if we remove the mass, the water will begin to flow again. We recognize that sunlight will cause paint to fade over time, that water will rot wood, that strong wind can damage a roof, and so forth. These are ordinary, everyday *causal facts* that we are familiar with; they are examples of the laws of physics and of causal determinism. They are the reason we protect surfaces from sunlight, paint fences with water-resistant chemicals, trim tree limbs periodically, and so on. We don't invent these laws, of course;

they are progressively discovered over the centuries by means of scientific research. Indeed, let us not forget that *they make science possible*, for it is only because causes and effects in the universe follow laws—that is, that nature follows *consistent patterns*—that we can formulate explanations of events and make predictions about what will happen with regard to future events. If there were no laws, there would be no science (another remarkable fact about our universe).

Let us take an example from a field nearer to our topic, biology, and examine it in the same way. I recently bought a plant at a home improvement store, and I no sooner had it home than it died! This despite the fact that it had an information card attached explaining how to take care of it and boldly proclaiming that it would bloom every year! The information on the card is based on a deterministic view of science, this time in the field of biology. Botanists studied these plants over an appropriate time frame and discovered progressively that they would live and thrive for several years, if maintained in a certain type of soil, with a certain amount of light, water, temperature, and so forth. They found that if you gave the plants too much water, or too little, they would die, and gradually determined the right amount (such as two ice cubes a week for orchids!). They also found that if room temperature goes above 75 degrees, certain plants will struggle, and if it is above 80, they will die, and so forth. Researchers reached many of these conclusions by performing experiments, but also discovered some no doubt by accident. Now, again, this is a deterministic process. The botanists might not know the exact way heat affects a plant—what it does to the makeup of a plant—but they do know that if there is too much, it kills the plant. They use this kind of knowledge to put together the information card that comes with the plant in a retail store. As we saw in the previous chapter, it is not quite possible to speak of "laws" in biology because of the complicated nature of the subject matter, but nevertheless, that is what researchers are modeling, and the probabilities they identify are based on the supposition that nature works by consistent laws. So, there is no chance involved in how the plant behaves, only unpredictability from our point of view. There is a cause for why my plant died, even if we can only guess at it. There is always, in short, a *causal reason* for a happening or event in our universe, even if we cannot always discern it with precision. This is the standard understanding of science. *It is based on our experience of the universe, not on prior assumptions about how things must work.*

Let us take a more sophisticated example from physics, from the work of Nobel-winning physicist Steven Weinberg.[2] Weinberg has written a fascinating book in cosmology, building a case for his view of the origin of the universe. He first argues for the Big Bang theory, that the universe began in a hot and dense state approximately fourteen billion years ago, and backs this up with evidence from astronomy and theoretical physics. He appeals to current cosmic background radiation, among other evidence, that was left over from the Big Bang to reason backward, as it were, to the conclusion that the Big Bang was the cause of this radiation. One of his main claims is that the first three minutes were crucial in the formation of the universe; in the first second or so, neutrons and protons formed (out of quarks and their environment); this led to nuclear reactions that produced hydrogen and helium. This process was followed by nuclear fusion reactions, along with the formation of helium-3, tritium, and lithium-7. (These early developments are responsible for the makeup of our universe today consisting of 75 percent hydrogen and 25 percent helium, according to Weinberg.) After further expansion and cooling in the first minutes, nuclear reactions stopped, and there were no heavier elements due to a lack of stable nuclei with atomic mass 5 or 8, and so the universe entered a dark age, full of helium, hydrogen, and radiation. The next step would be the formation of stars and galaxies.

This brief description of Weinberg's fascinating but technical account of how the universe formed is enough to give us non-physicists a flavor of his argument. What we need to note is that there is no reliance upon or even clear mention of the notion of chance in his logical reconstruction of the formation of the early universe. Weinberg takes present data, and, along with the laws of physics and mathematics, reasons backward to what *must* have occurred at the beginning of the universe to produce that current data. There is no room for chance anywhere in this explanation. His account has the form, and must have the form, E was caused by D, which was caused by C, which in turn was caused by B, all the way back to A. This process must be understood in a deterministic way, which is the way Weinberg understands it, though he does not discuss the notion. (And we should note that he was suspicious of indeterministic accounts of quantum mechanics.) Otherwise, we could not conclude that for C to occur, B must have caused it (or, more accurately, that B must lead to C; for example, quarks lead to the formation of neutrons and protons).

2. See Weinberg, *First Three Minutes*, 101–21.

If we were to allow that chance was involved in this process, we could not reason from the formation of neutrons to quarks as their cause. Any uncertainty in Weinberg's conclusions is an epistemic uncertainty and not an ontological indeterminacy in nature, a notion that he resisted.[3]

Two more brief cases from biology will serve to drive our point home. The first concerns *mitosis*, a fundamental process for life. We recall from Biology 101 that during mitosis, a cell duplicates all of its contents, including its chromosomes, and then splits to form two identical daughter cells. Because this process is so critical, the steps of mitosis are carefully controlled by certain genes. When mitosis is not regulated correctly, health problems such as cancer can result. Now, this brief description is deterministic in nature, with the prior steps understood as leading by necessity to the later steps (all other variables being equal). We study the process in order to discover the deterministic steps, and the more we know, the more control we have to prevent problems (which also occur in a deterministic way). A second example is the vexed question of *the evolution of elephant trunks*! Although there is much speculation about how the trunk evolved, one current theory suggests that climate changes had an effect on the grasses the elephants used for food; this in turn affected the way elephants used their jaws and their trunks, and, over time, these actions changed the nature of the trunk, which gradually became longer and more flexible.[4] Although one can see that this valiant reconstruction is far from convincing, it is one of those many common explanations in evolution that appeals in part to animals' interaction with their immediate environment. It is also yet another example of how evolutionary biologists do not pay enough attention to the deterministic nature of this process. For the simple fact is there is no part of this causal story about the possible evolution of elephant trunks that involves any chance happenings. The changes in the climate, the effects of the plants and

3. It is sometimes claimed that chaos theory may provide evidence of an indeterminism in nature, with appeal to the so-called "butterfly effect." The main claim is that small changes can lead to large, unpredictable effects over time. While this may be true, it does not succeed in showing an indeterminism in nature, for as Wildman and Russell point out in a detailed mathematical analysis, the butterfly effect actually supports metaphysical determinism. This is because it testifies to the "high degree of causal connectedness in certain natural systems"; they further add that, "we can say without hesitation that chaos in nature gives no evidence of any metaphysical openness in nature" (Wildman and Russell, "Chaos," 82). Indeed, logically, the suggestion that a butterfly flapping its wings may begin a causal chain of reactions that results in a hurricane in another part of the world requires a *deterministic* universe to make any sense.

4. See the account in Quan, "Scientists."

grasses on the trunks over time—all of these causal happenings follow a deterministic path. No part of the process, as far as we can see, occurs by chance. (And if one were to claim that the elephant's wanderings are a matter of chance, this assumes a position on an open question—whether animal behavior is or is not deterministic; similarly, if one claims that changes in the climate were a matter of chance, or that the soil the grass grows in just happened to be especially susceptible to climates changes—all of these claims simply beg the question, and remind us of the generally cavalier way that evolutionary biologists treat the vital issues of chance and determinism.)

These everyday examples illustrate what it means to say that science operates deterministically and, speaking somewhat loosely, that we live in a deterministic universe. However, it is essential to keep this kind of causal determinism distinct from what is sometimes referred to as the *causal closure* of the universe, two notions that are frequently and mistakenly conflated with each other by a number of thinkers.[5] Some thinkers, it seems, without discussion and under the influence of a contentious reading of an idea from Pierre Laplace (1749–1827),[6] just assume that determinism means causal closure, or perhaps we should say that they identify the world of Newtonian mechanics with a closed system (seemingly forgetting that Newton himself did not understand it this way).[7] While I affirm determinism about the way the universe operates, which is uncovered in scientific research, I deny the causal closure of the universe. The causal closure of the universe is sloppily presented as the view that if the universe is deterministic, then there can only be *physical* causes in the universe, with no interference from outside (and so it is said to be "causally closed"). It is often (erroneously) characterized as the view that whenever there is a causal event, it *must* be a physical causal event, and there can be no other kinds of cause. This understanding is attributed to Laplace who suggested that if there was a demon who knew the exact state of the universe at any given time and if the universe operates deterministically, then we could predict the exact state at any future time, because the deterministic nature of causation would mean that every future event had to happen, given the laws of science and the initial starting point.

5. On the notion of causal closure, see Plantinga, *Where the Conflict Really Lies*, 70–90; Koperski, *Divine Action*, 120–23, 144–45.

6. See Laplace, *Philosophical Essay on Probabilities*.

7. For a helpful discussion of Newton's views on determinism and God's action, see McDonald and Tro, "In Defense of Methodological Naturalism," 201–29.

Laplace's view is often interpreted to entail the denial of human free will and any action by God in the world, because these would be non-physical causes, which are precluded by determinism. However, this is a fallacious move.[8] To see why, we must distinguish between determinism and the causal closure of the universe.

Laplace's claim is true only if we make two assumptions: (1) that God does not intervene in creation; (2) that human beings do not have free will and so cannot freely affect happenings in nature. However, we do not need to make these assumptions to hold to a deterministic view of science and the operation of the physical universe, and I deny both claims. My position is that determinism holds *unless* (or until) God intervenes, or *unless* (or until) human beings, with their free will, intervene, as I indicated in the previous chapter. (I hold, along with many others, that determinism is not a threat to free will—I am not a materialist about the conscious mind, nor does evolution establish any kind of materialism or emergentism about the mind.)[9] *But otherwise the universe operates deterministically*. So, if God intervenes to cause a particle to move one way rather than another (a topic we will discuss in detail in chapter 6), the particle operates deterministically *up to the time God intervenes* and *after he intervenes*. So in this sense, the universe, in my view, is not causally closed. Similarly, if a disease has taken hold of a person, and doctors freely give the patient a drug to kill the disease, the person's body operates deterministically both *before* and *after* this action (until the next human intervention). This, I claim, is how our universe works; *when left to its own devices, the world operates deterministically*. Human beings are not subject to causal determinism because we have free will, and we are free to disrupt, intervene in, or alter causal chains that otherwise operate

8. See Clayton, *God and Contemporary Science*; Clayton claims, for example, that if Laplace's approach is correct and a strict determinism would allow us to predict all future events, and retrodict all past ones, then there would truly be no place for divine (or free human) action (see 206). Wildman makes the same claim (see "Chaos," 84). This mistaken reasoning is typical in this literature. As I will suggest, God could still intervene in a deterministic system, and so could we; so free divine and human actions would still be possible. It is true that it would not then be possible to make fully accurate predictions in some cases, but prediction, although conceptually related to the notion of determinism, is not essential to it if you allow for outside intervention. These thinkers often talk as if a deterministically operating universe excludes free actions and even God intervening, when this is not true (as we will see momentarily). If one is going to reject determinism in the physical universe, this is *not* a reason to reject it.

9. See Murphy, *Bodies and Souls*, for discussion of a variety of views on the topic and for her defense of a form of emergentism.

deterministically. Indeed, we rely on their operating in a deterministic way when we intervene, as our discussion of how science works has shown. (Animals, by contrast, are subject to causal determinism, on my view.) The same applies to any intervention by God in creation. Because God and especially humans can intervene, it may compromise our scientific predictions in certain (very) limited cases, but the universe in itself still operates in a deterministic way. Nor can we say that a deterministically operating universe restricts or confines God, since he planned it all out, and can intervene whenever he wishes!

Sometimes, an objection is raised that if God acted specially in creation, there would be no further deterministic effects after that, or there would be no way to make predictions. But this is not true for several reasons. One, a small intervention would not upset the course of the universe in any noticeable way, and it would not upset or compromise the law-like behavior at all from our epistemic vantage point. If God intervenes, the laws operate in a fully deterministic way from the point of his intervention onwards. Two, we already see this kind of intervention from free human beings, and we can still make very accurate predictions (even in quantum mechanics, where the intervention is supposed to have an effect on the result). It is a fallacious move to think that if the universe operates deterministically, then it follows that there is no free will or that God either cannot or does not intervene in his creation. Yet this is a general fallacy that seems common in contemporary theology. Some theologians seem to think that if you say the universe is deterministic, then you are saying that it is causally closed (and so are denying free will). Looking at the matter from the other way round, it is fallacious to claim that if we cannot predict the future or reconstruct the past, then nature must not operate in a deterministic manner.

There are two concepts present in Laplace's thought experiment: determinism and predictability. The first is the main concept, however, not the second, because even if we have intervention from either God or humans in specific cases that make predictability impossible, determinism about the way nature operates still holds, as does his general point in principle. In short, we should not identify the causal closure of the universe with determinism, since they are clearly distinct notions. It is important to keep them distinct to avoid confusion, and indeed, very few theists hold that the universe is causally closed.

Strictly speaking, then, my position is that the universe is partially deterministic, since God and human beings can affect the future outcomes. Predictive failures, therefore, are due either to the complexity of the causal chains or to intervention by God or humans, but not because of chance. From a scientific point of view, the universe is fully deterministic, so, with this clarification, I shall continue to describe my view as determinism.

Laws of Nature, Freedom, and Contingency

In order to appreciate the significance of this deterministic view of the universe, we need to highlight some additional points. First, let us remind ourselves of the remarkable fact that the functions of not just the earth but also the whole universe are governed by what we refer to as scientific law, or sometimes as the laws of nature, the laws of science, the laws of physics and chemistry, or occasionally even as natural laws. What are these laws? They are "regularities" or "patterns" in nature that govern how things behave. We discover these laws as we practice science, because we find over time that nature behaves in uniform ways. Such laws include the laws of physics and chemistry, perhaps the two most basic sciences (along with mathematics) for understanding the physical universe. Examples would include Newton's laws of motion (his second law states that the time rate of change of the momentum of a body is equal in both magnitude and direction to the force imposed on it); Maxwell's laws of electricity; chemical laws; as well as many others that we remember from our school textbooks, and that scientists rely on in their everyday work. Indeed, we all appeal to them informally in our day-to-day activities, often without realizing that we are doing so, as we have seen above. We have all discovered, non-scientists and scientists alike, that this is how the universe works. Science under one aspect can be regarded as a formal, very sophisticated way of thinking about these laws and applying them in our understanding of physical reality. Think of the application of these laws, for instance, in sending a rocket ship to the moon. Our knowledge of the way the physical realm works allows us to calculate how much power or thrust we need to escape Earth's orbit, how warm the rocket should be to protect the astronauts against the cold of space, how a rocket engine needs to be constructed to ignite on the moon, how to maneuver the rocket in the forces of space. It is because nature follows consistent patterns in how it behaves that we can work out all of these things using our knowledge of causal laws, past inductive evidence, experiments and tests, mathematics, and ordinary logical reasoning.

Second, it is important to note that the laws of science are arrived at inductively—that is, we observe nature behaving in consistent patterns in all instances of the same type, and we conclude that it will behave the same way in the future. So we accept the conclusion that there are no exceptions to the laws, that, all things being equal in an experiment or a process, we know *with certainty* what will happen. If it fails to happen, we do not conclude that it was because in this one instance the laws failed to hold (believing that they hold in most instances but not all). No, we conclude it was because all things were not equal (or quite accurate) in the experiment, that something went awry in the setup or the ingredients that prevented the expected result from occurring. So when a car fails to start, mechanics do not conclude that the electrical law governing explosions does not always hold. Instead, they begin looking for an air or a fuel leak! When they find and fix the leak, the car will start! If it does not, they will conclude that they had mistakenly identified the cause and begin to seek another, but will *never* entertain the view that the problem is that the laws do not hold, or that real chance plays a role in the working of the laws in the engine!

The question of the ontological status of the laws of science has been the subject of some discussion. Do the laws exist in themselves as independent entities in the universe (a Platonist position), or are they best understood as descriptions of the way that objects in the universe behave? As we have seen, they behave in "regular" or consistent ways to the extent that we can describe these patterns as "laws." Some thinkers, like Hume, were skeptical about whether the necessity that is the basis of scientific law is actually a feature of reality or whether it is mostly an imposition by the human mind (a minority view, though there are a variety of positions on the nature of scientific laws, and no agreement on how they are best understood). Indeed, it was not until after Newton that philosophers started to give this topic serious thought, when it became clear, as science developed, that scientific law was a crucial feature of the universe. Medieval thinkers did not give much attention to the phenomenon, since they tended to understand causation in terms of Aristotle's four causes. René Descartes seems to have been one of the first philosophers to give serious consideration to the topic, giving the notions of regularity and mechanism a prominent place in his understanding of the universe. The later developments of the corpuscular theory and atomic theory also led to sharp focus on the status of the laws. Although the debate is far from settled about their ontological status, there is general agreement that such laws are the basis of science in the sense that they reveal regularities in nature (which medieval thinkers would have said comes from the essence or nature of the way things behave in themselves). They reveal an underlying causal order, one that allows us not only to provide explanations for why things occur but also to alter causal happenings and to make predictions about what will happen. So the laws quickly become the underlying foundation of the scientific method. Fortunately, this debate is not crucial for our arguments here; the metaphysical status of the laws is less important than their functional role in science. All we need to affirm is that there are such laws, that they impose necessity on nature, and that there are no exceptions to them in the normal operations of science.[10] What matters from the point of view of the question of divine action is that the laws exist, that the physical stuff of the universe behaves in a certain way, so giving rise to the question of the role that God plays. (The various disputed theories about the status of laws just show that we are all trying to understand—inevitably unsuccessfully—what causation is, how it occurs, and, for the theist, ways in which God may be involved.)

Third, we should emphasize again the crucial role of free human actions in the face of determinism. We will define free will as the ability to make a reasoned choice between two genuine alternatives, a choice that is outside the realm of scientific causation. It is a remarkable feature of reality that beings have evolved on the earth that have this quality, *Homo sapiens*. One might think that since we have free will, we could not live in a purely deterministic universe. This is because we have the ability to freely intervene in causal sequences and so bring about (or prevent) things from happening that otherwise would not (or would have) happened. However, the existence of free will in human beings means only that *when we come along*, we are able to alter the causal sequences that are occurring deterministically in nature. This is not quite the same thing as chance since there is intention usually behind what we do. We noted several examples in the previous chapter from Aristotle. Happenings involving human beings and their free decision-making are often described

10. For a helpful account of various views, see Koperski, *Divine Action*, 86–107; also, Cartwright, *How the Laws of Physics Lie*; Hoefer, "Causal Determinism"; Silva, "Thomas Aquinas and some Neo-Thomists." We note from this literature that, although we all seem to operate with a similar understanding of how laws work in the practice of science, there is no agreement about their ontological status, yet science gets along quite well even without resolving this matter.

as "chance" events in the sense that they are not the result of a causally determined process (the essence of free will is that it is not subject to a scientific causal analysis). However, free human actions are not events involving real chance in the sense in which the term is meant in our discussion in the previous chapter. We accept that interruptions in nature, interventions from us, human free will changing the course of the history of the earth and events on the earth (but not in neighboring galaxies, or elsewhere in the universe) *occur*. In this sense, the universe is no longer fully deterministic *after the arrival of human beings*—because we can and do intervene regularly in the causal chains, something that was not possible before we came on the scene (though God could intervene in a direct way, of course). We also assume that lower-level animals do not have free will and so are subject to causal determinism, not just in how things affect them, but also with regard to their own behavior. (Indeed, sometimes we even suggest that human behavior is predictable in somewhat law-like ways, e.g., in game theory, consumer spending habits, investing habits, etc., attempting to apply the deterministic model of science to human actions, albeit in a flawed, incomplete way that is frustrated by the fact that we have free will and so do not behave deterministically by nature!)

So given that human beings have free will, we are not therefore subject to causal determinism (though our bodies are, as is their interaction for the most part with the rest of nature; nevertheless, we therefore reject a *totally* mechanistic account of reality). If one denies that we have free will, then one must hold that even human beings are fully subject to determinism, just like everything else in the universe (assuming one is a practical determinist, which most people are, including most scientists). This is a very difficult position to maintain with consistency, for even those who deny that we really have free will act on the assumption that we do! (Just as those who deny determinism act and live on the assumption that it is true.) I believe consideration of the topic of global warming illustrates these points clearly. Within this debate, there is the *scientific* argument, which asserts that global warming is occurring, that if we do not take urgent steps to address the problem, it will get much worse, perhaps irreversibly so. But there is also the *moral* argument, which is the claim that it is immoral not to take action soon, since the effects of global warming will be catastrophic for the human race and for the planet; failure to act to prevent this is morally wrong. The appeal in this judgment and plea is founded not just on a realm of objective moral values but also on the belief that human beings have free will. And so if we do not act to prevent global warming, we are doing something morally wrong for which we can be held responsible. This claim is based on the belief that we are free to choose to curtail the use of fossil fuels, to switch to electric cars, and to adopt other energy-saving measures. If we do these things, it will have a positive causal effect, so goes the scientific argument, on the climate. This scientific prediction is based not just on our knowledge of causes and effects in this area, *but also on an assumption of causal determinism*. It is based on an argument that allows no room for chance in the pathway between our current state and the future state that will occur if we do not take additional steps now to alter the causal pathway. There may be unknown and unforeseen variables, as we noted above in our examples, but these are not cases of chance. We also see how human intervention by means of free will is not chance either, though it does have the effect of changing the course of causal chains that would have gone some other way if we had not intervened. In the case of global warming, the scientific question assumes determinism and the moral question assumes free will.[11]

Our discussion so far brings us more directly to the philosophical or metaphysical question concerning the nature of physical reality, whether it is deterministic or indeterministic in itself. I have been arguing that it is deterministic (we will consider the objection to this view from quantum mechanics in our next section). But it is important to emphasize that, as theoretical physicist and philosopher of science James Cushing notes, science itself does not answer the metaphysical question.[12] So one must therefore make a philosophical decision, and indeed, we must acknowledge honestly that in considering these deeper issues, all of us make metaphysical commitments that cannot be fully justified by either logical or scientific reasoning. It seems that physicists can only say whether a physical system is predictable or not (sometimes referred to as epistemic indeterminism, an imprecision based on our lack of knowledge but not

11. We must keep always in mind that statistical analyses that rely on several random variables (such as calculating the average height of ocean waves in a harbor during a certain time of year) report only our epistemic predictions about ontological realities, but what happens *in reality itself* (in the ocean) is not random. We should also be careful to keep distinct those statistical analyses that appeal to human behavior, and so freewill, which is outside determinism (such as how many people are likely to benefit from a new exercise treatment for arthritis), and those that do not, which are subject to determinism (such as in our ocean example). These confusions are endemic in the literature on religion and evolution (for example in explanations of the process of genetic drift).

12. See Cushing, "Determinism Versus Indeterminism."

necessarily one that exists in reality itself). But if they go further and assert indeterminism about reality (ontological indeterminism), they are then making a metaphysical (or a philosophical) claim. It seems, however, that the scientific evidence is compatible with either view. Indeed, some thinkers suggest that because of the difficult nature of the subject matter, especially at the quantum level, the question may be ultimately undecidable.[13] Whether or not that suggestion is correct, and despite the often fractious rivalry between classical mechanics, quantum mechanics, and relativity theory, for now the question of determinism/indeterminism remains open. And we should not forget that the apparent anomalies that arise in and between these theoretical approaches are themselves all the subject of intense debate.[14] Despite all of this, working, functional science proceeds based on an assumption of determinism.

In order to help us with the choice between the two, I suggest we keep in mind the principle of sufficient reason. Attributed to Leibniz, this principle is based on the simple logical point that for every event or happening there must be a cause or explanation for its existence, a reason for why it is the way it is and not otherwise. Leibniz and many others noted that this would seem to be a basic law of logic and common sense, especially when applied to ordinary causal happenings in nature. However, an indeterministic interpretation of quantum mechanics, in particular, would seem to challenge the principle of sufficient reason because it suggests that an event can occur at the quantum level, which, even though it falls within a certain measurable range, does not seem to have any specific cause. Of course, many hold that this interpretation is unintelligible, in part because of the principle of sufficient reason, and that it makes more sense to say that the quantum events have definite causes that are unknown to us, an incompleteness that arises because of our lack of understanding (whether an intractable problem or not). (So another challenge for those theological approaches that support divine action as taking place in a world of indeterminacy and contingency will be how to reconcile quantum effects with the principle of sufficient reason.)

A deterministic understanding of reality is a direct challenge to the notion of contingency that we find in the work of Stephen J. Gould, a

13. For more on this point, see Suppes, "Transcendental Character of Determinism," 242–57; Suppes argues that "Deterministic metaphysicians can comfortably hold to their view knowing they cannot be empirically refuted, but so can indeterministic ones as well" (254).

14. For a not too technical overview, see Koperski, *Divine Action*, ch. 6.

concept that has had an almost defining influence on how the theory of evolution is understood and taught today. We referred briefly to Gould's famous and very influential argument in the last chapter, that if we replayed history again, "the tape of life," things would go differently the second time around, and we would get different results, including different species, and almost certainly no species of *Homo sapiens*.[15] This is because of the operation of (real) chance, which affects the process of evolution. However, if we are correct that the universe operates in a deterministic manner, then Gould's conclusion is misguided. This is because on the deterministic view, there is no chance involved (and keep in mind that Gould is not just talking about the domains of biology and evolution, he is talking about how the underlying physics of the universe operates). It follows from this that the universe would go the same way a second time around *if it were set up with the exact same initial conditions as the first time around*. If the initial setup is the same—exact same ingredients, same laws, time line, everything identical—we would get the same results (at least until the appearance of free human beings, and assuming also no outside intervention by God or anything else). This is how a deterministic universe works. Identical causes lead to identical results, as long as all other things hold constant.[16] Only if the initial ingredients were different would we get different outcomes. Gould's view is based on a (metaphysical) assumption about reality, namely, that it is indeterministic and operates with a degree of chance.[17] One wonders, of course, if he held to the same view consistently in physics. It seems to me to be common for many to hold that physics is deterministic but that biology is not, an inconsistency at the heart of positions like Gould's. Logically speaking, it seems quite plain that you cannot have chance and contingency in a deterministic universe. And the same seems to go for purpose as well, since if a causal chain is subject to chance, and so could go in a number of different ways to the way it ends up going, this is difficult to square with the view that a particular pathway was intended.

There seems to be, if not a contradiction, then a serious tension between committing to the view that the universe is indeterministic

15. See Gould, *Wonderful Life*, 320–21.

16. On this point, see Shanahan, "Evolution Indeterminism Thesis," who notes that the debate over determinism/indeterminism "is now one of the liveliest topics in the philosophy of biology" (163).

17. For a helpful discussion of these points, see Beatty, "What Are Narratives Good For?"

and significantly governed by chance, and yet claiming that modern scientific evidence makes it difficult to believe in exceptions to scientific laws. A commitment to indeterminism means that events are not fully determined by scientific laws. So indeterminism introduces an element of chance into the universe, and would seem to place some limitations on the universal nature of scientific laws. It seems that if you stick to the view that scientific laws always hold, you should be a determinist about reality (at least functionally, if not metaphysically). We can see then that the deterministic view throws a fresh light on the key question about how the universe was set up initially and whether it was set up for a particular purpose. A deterministic understanding, as we will see, suggests a particular answer to this question. This is because determinism can play a key role in an argument for design or purpose, *given the order and complexity in the universe.*

Indeterminism and Objections to Determinism

There are objections against determinism in general as a scientific understanding of how the physical universe behaves. Some of these objections appeal to science (especially when we consider the theory of quantum mechanics); some appeal to philosophy and metaphysics, where some thinkers advance an indeterministic view, though (as we have seen) the question is still very much open. There are also objections to determinism from a theological point of view, relating to the way in which determinism would affect our understanding of God's action in the world and how it might relate to the process of evolution (our main concern in this book). We will address the scientific and philosophical objections here, and the theological objections in later chapters.

There are a few commonsense-type objections to my argument that the universe follows a deterministic path that we should first consider, before we look at more sophisticated worries. One commonsense concern might be that some things in our world do seem to happen by chance, even though we believe that there is generally some kind of deterministic causal explanation lurking as well. On the surface, it might seem counterintuitive to say that there is never any chance involved in the occurrence of natural events. However, a few moments reflection will show that this is a superficial view because determinism is how we normally think about science. We only think about chance at an abstract level; at the concrete level, we work as if it does not exist. Suppose, for instance, that a plant gets a disease because there "just happened to be" an animal seeking food that carried the disease to the plant when wandering across the plains, or suppose that there "just happened to be" a strong wind on a particular day that blew infected debris onto the plant, thus passing on the disease. Although cases like these might seem on the surface to be chance happenings, it is important to realize that this is not so, and that the phrase "just happened to be" is not the best description of what has occurred. This use of the term "chance" might fit into the psychological sense of chance we identified in the previous chapter, and could also be described as apparent (but not real) chance. This is because there is a deterministic causal process that explains why the animal is a carrier of the disease and why it visited the plant and spread the disease (assuming that the animal does not have free will). Similarly, there exists a full causal explanation for why the wind is blowing where it is blowing, its strength, trajectory, timeline, and its effects. This is how science understands the operation of nature (indeed, the science of meteorology is an attempt to discover these kinds of causal chains in order to make more accurate weather predictions, predictions that would not be reliable unless determinism were true). So neither of these happenings is a case of real chance, nor do we regard them as such when we actually begin to study the type of causation involved and whether we can influence it in order to prevent undesirable events.

This point also holds true for causal chains that interact with each other, which is often the basis of a second objection—that a causal chain might in itself be deterministic but its interaction with other, separate and distinct, causal chains is a matter of chance. But in fact, the above examples involve the interaction of causal chains, and so our response is the same. In all such cases, there is a scientific explanation for each causal chain and its interaction with other chains. So there are, in short, no chance or coincidental meetings of causal chains. When we do science, we assume this in our work because we assume two things: that the laws of physics always hold (no exceptions) and that the deterministic nature of causation is upheld in the universe. We do *not* assume that chance occurs anywhere in nature's unfolding; that is to say, let us recall, that some laws occasionally do not obtain or apply, or that A (all other things being equal) does not always lead to B (e.g., a water force of X will destroy any building whose foundational support is less than X, all other things being equal, which means that we have done our calculations correctly, have discerned which laws are involved and know how they work with certainty, have allowed for all other causal factors that would or might affect the outcome, and so forth). The interactions of otherwise distinct causal sequences, therefore, do not include any elements of chance (even though the sheer unpredictability of various events often tempts us to speak loosely as if chance is present, and even though from a psychological point of view it may be difficult to avoid describing such events as "occurring by chance"). So this objection does not take full account of the fact that each effect in a causal chain is itself determined by prior causes, and so the whole causal chain *must follow the path it does follow*.

As we noted in the previous chapter, Darwin was ambivalent about the notion of chance and often used it in different ways. He writes: "Throw up a handful of feathers, and all fall to the ground according to definite laws; but how simple is the problem of where each shall fall compared with problems in the evolution of species."[18] Here, he seems to recognize that the feathers fall to the ground according to a deterministic process, but fails then to extend the point to draw the same conclusion about the way evolution itself operates. We must recognize that there is no coming together of causal chains by coincidence on this deterministic view. So we are led back then to consider the *beginning ingredients* out of which the initial causal chains emerge and how they got that way. The fact of the existence of this beginning and its makeup and what it leads to *now takes on an enormous significance*. It raises the question: *How was the universe set up, by whom, and, by extension, for what purpose?*

There is a more sophisticated objection to a deterministic view of the universe, one that comes from the theory of quantum mechanics. Unlike classical physics, which studies the ordinary objects of our experience such as volcanoes, rivers, planets, living things, forces, energy, and electricity at the macroscopic level, quantum mechanics is an area of scientific research that focuses on the nature and interactions of subatomic and atomic particles, such as atoms, quanta, neutrons, electrons, quarks, photons, and so forth. Particle physicists study the energy and momentum of these particles, the laws by which they operate, and their relationship to the macroscopic world. Some of their findings are unusual and raise questions relating to metaphysical topics concerning determinism and indeterminism and the applicability of scientific law. The quantum world also raises questions about the distinction between reality itself

18. Darwin, *Origin of Species*, 77.

and our knowledge of it (where predictability and unpredictability are relevant), as well as about the notions of chance and necessity. Although from the practical side, quantum mechanics has been very successful in areas like computing, transistors, lasers, modern electronics, nuclear energy, and many medical applications, it continues to stir debate at the theoretical level, with several theories being proposed concerning what we can conclude about nature on the basis of our research into this unusual area.[19]

Researchers have discovered that we run into interesting conundrums when studying subatomic particles, difficulties that do not arise in the ordinary macroscopic world of our experience. One major difficulty is that quantum effects seem to be affected by the methods used to study them, which means that it does not seem possible to achieve a truly objective understanding of the nature of the microscopic particles. It seems that when we are studying the particle as observers, we are also changing it, affecting how it behaves. Research suggests that particles/wave functions don't seem to have any definite properties before scientists start trying to measure them, at least none that we can detect and measure. When we measure the system, it takes on a definite value, and so the act of measuring or the interference of the observer seems to be forcing structure and discreteness on a reality that does not have it prior to that. This applies to instances of quantum effects such as radioactive decay, photon emission, and absorption, where precision does not seem to be possible, only predictions that fall within a certain range. Walter Heisenberg proposed the uncertainty principle, which says that we cannot know both the position and the rate of motion of a particle at the same time.[20] An important consequence of this state of uncertainty is that we can't state the laws concerning the behavior of particles operating at the subatomic level with precision; they have to be expressed as statistical probabilities. So in this way they differ from the laws governing macroscopic objects, which, although arrived at by induction, are nevertheless accepted as laws in the sense of holding in every given instance, and so can be stated precisely.

However, there is disagreement over the correct philosophical or metaphysical position to take with regard to what is really going on at the subatomic level. Heisenberg and Bohr favored a theory of indeterminism

19. For a good overview of quantum theory, including the various interpretative problems, see Polkinghorne, *Quantum Theory*.

20. See Heisenberg, *Physics and Philosophy*.

about reality. Indeterminism is usually thought of in terms of what it denies (namely, determinism!), holding that, given a set of initial conditions together with the laws of physics, no future outcome *has* to occur; the outcome is not certain. So the future event that occurs from the initial conditions is not predictable, even in principle. With regard to the notion of chance, the indeterministic view is committed to saying that there is real chance in nature (and not merely unpredictability). Chance, we recall, is the notion that given a prior cause, the effect does not have to occur (because the law does not always hold). An event might be said to be *probable* in some cases, as in quantum mechanics, but *not definite*, even in principle. We are reduced to probability on this view not because there are hidden variables that we are unable to discover at present or because the causal field is too complex, but because *reality itself is indeterminate*; that is, it does not follow causal laws without exception. So one consequence of an indeterminist view is that it allows for an element of chance in how nature operates, whereas determinism does not.

Yet the indeterministic interpretation of quantum theory is only one possible interpretation. The main question is whether the indeterminacy we observe, and that frustrates us in our attempts to state the laws that govern subatomic particles in precise terms, exists *in the particles themselves*, or is only a result of the interaction of scientists as observers with the particles, an interaction that affects their scientific results. Einstein found Heisenberg's interpretation counterintuitive and, indeed, unacceptable. It was also rejected in favor of a deterministic interpretation by Max Planck and more recently by physicists such as David Bohm.[21] These thinkers and others argue that there are hidden variables that determine the outcomes, just as there are in many macroscopic causal systems. (Some theologians suggest that God might be the hidden variable—we will consider this view in the next chapter.) It is very difficult to make a judgment here about the metaphysical question, but there are three very

21. Bohm, *Causality and Chance in Modern Physics*, 64–65, makes this observation about the debate: "For it is quite possible that while the quantum theory, and with it the indeterminacy principle, are valid to a very high degree of approximation in a certain domain, they both cease to have relevance in new domains below that in which the current theory is applicable. Thus, the conclusion that there is no deeper level of causally determined motion is just a piece of circular reasoning, since it will follow only if we assume beforehand that no such level exists. . . . [But it] follows neither from the experimental facts underlying the quantum mechanics nor from the mathematical equations in terms of which the theory is expressed."

important points that can be raised in favor of the deterministic thesis we are interested in this book.

The first is that at the practical level of doing science, determinism is the standard view, as we pointed out above, no matter what metaphysical position we take with regard to quantum mechanics.

The second is that quantum effects do not usually carry over to the macroscopic world, although there seem to be some exceptions, and there is no agreement among thinkers on the issue. Indeed, these effects seem not even to be all that significant at the quantum level itself; otherwise, we could hardly build lasers and computers based on an indeterministic and unpredictable quicksand. So not only is quantum mechanics itself mostly deterministic, but the quantum effects appear to disappear or are insignificant when they interact with larger numbers of atoms and molecules in the macroscopic world and also in the larger environment. This seems to be especially true in biology, and in evolution, one of our main concerns here. So this would mean then that the process of evolution is subject to the deterministic nature of the macroscopic world (despite quantum mechanics).

The third point is the one raised by Einstein when he observed that "God does not play dice with the universe," a remark that suggests there cannot be this kind of uncertainty and indeterminacy, with events subject to real chance, present in nature itself.[22] The point, I believe, behind Einstein's suggestion is simply that it is not clear whether the notion of an indeterminacy in nature itself even makes any sense. It seems to be something that is not possible logically, and if not, then it does not seem to be an intelligible notion. The idea that subatomic particles, or aspects of particles, have no determinate nature until we interact with them seems unintelligible on its face. Of course, we should not be too quick to rule it out, since reality is strange and full of surprises. However, this is one idea that might just stretch our intuitions too far. My view is that the thesis that nature is deterministic is the most plausible one to adopt on the grounds that all of our experience is against the conclusion that the behaviors of particles could be indeterminate in themselves. But it seems clear that science itself does not settle the question and that one's attitude or position on these deeper issues hinges on the philosophical decision one makes about the metaphysical question regarding whether

22. See Einstein's letter to Max Born in *Born-Einstein Letters*, xxii.

determinism or indeterminism is the most intelligible position logically, and the most in line with our scientific understanding of the world.[23]

Chance, Randomness, and Biology Revisited

We pointed out in the previous chapter that evolutionary biologists usually wish to deny that there is any purpose in the process of evolution. They sometimes express this point by saying that the process is not aimed at producing any particular species, or that there is no mechanism in nature that drives evolution toward certain (particularly favorable) goals or outcomes. More specifically, they hold the drivers of evolutionary change, the nature of the environment in which the process takes place, and the changes to DNA on the side of the organism are not directed in any way, but occur by chance. And so this means that the outcomes of the process, which over time leads to an array of increasingly complex species, including the most advanced one, *Homo sapiens*, are a matter of chance. None of these outcomes had to occur, despite appearances to the contrary (that is, that such complexity speaks against chance playing much of or any role in the process, and that it is counterintuitive that the design we see in nature—in the complexity of organs and whole species, and of fit with the environment—is merely apparent, not real).

The notions of chance and randomness, though distinct, are intimately related and are best understood—at least in any conversation concerning evolution—in relation to each other.[24] It is not only biologists who need to be careful when describing the occurrence of events in nature. We don't want to say that because an event occurs by chance, that it therefore has no cause. We know that this is not true, and indeed we can usually point to the cause (or likely cause) of any particular mutation, or environmental change, as our knowledge of causes and effects in the process of evolution increases, helping us to make more accurate predictions and suggest promising lines of research. Yet we saw in the last chapter that evolutionary biologist Ayala seems to run the notions of chance and randomness together. He sometimes uses the terms synonymously, as do many others. We noted that it can be helpful to distinguish two senses of

23. This point is very clear I believe when we consider Schrodinger's cat thought experiment designed to show the absurdity of the Copenhagen interpretation when applied at least to macroscopic objects; see Polkinghorne, *Quantum Theory*, 51–52.

24. See my *Evolution, Chance, and God*, 106–13, for further analysis of these concepts.

randomness—one where it refers to an event that occurs by real chance, and the other that refers to the sense that the process of evolution has no teleology, is not aimed at bringing about any particular end with regard to the development of an organism. The end or outcome that occurs is random. Perhaps it is better to describe the first meaning as "chance," and keep the word "random" for the second meaning, which better reflects what it is claimed is going on in evolution.

Let us bring out more clearly the difference between the two meanings, but also the intimate connection between them. I like to define these key terms in the following way:

> **Chance:** *To say that an event came about by chance means that the prior cause(s) of the event could have been otherwise, and so the event might not have happened* (in the way explained in the previous chapter).
>
> **Randomness:** *To say that an event is random means that there is no goal or purpose or end toward which a chain of causes and effects is striving (for example, in the development of an organism), and which the various causes (for example, in mutations) are aimed at bringing about, or to which they are contributing.*

These two definitions keep the terms clear. Claims about chance are about the causes of evolutionary change, specifically that the causes are not necessary. Claims about randomness refer to the *outcome* of a causal process in evolution (specifically that the end result of the process was not intended, directed, etc., either by a mind or by a mechanism in the physical process that led to the end result).

In everyday usage, we often use the term "chance" when we mean "random" (no intended purpose) and also "random" to mean "chance" (when we mean that the cause could have been otherwise), but in our analysis, it creates less confusion overall if we try to mostly keep these respective meanings of the terms distinct, rather than using the terms interchangeably.

The two notions, however, though distinct, are inextricably linked because there seems to be a clear relationship between chance and purpose (teleology). This is because *the cause(s) of an event, and by extension the causes of a chain of events, determine the outcome or the goal of the chain.* This is how science works in practice, an illustration of the deterministic nature of science. But if one insists on the presence of chance (i.e., that the cause of a particular event is not necessary, is not determined)—so the event need not have occurred—then it follows that the end result does not have to occur either (or, more accurately, that no particular end result is intended, the result occurs randomly).

So if there are elements of chance in a causal series, then it is hard to see how the outcome can be any specific purpose that is aimed at from the beginning of the series. This is the position of the atheistic naturalist. However, some theists deny that this conclusion follows, because they believe that it is logically possible for God to put some chance into the universe but still bring about certain intended ends.[25]

Thinking about determinism and its relation to chance prompts two questions: (1) Could God create a world that in some way operates with elements of *both* determinism *and* chance? (2) Could we have a physical world in which determinism is true, but there are also real chance events? (The second question restricts the query to the domain of science.) Although the answer to these questions surely must be "no," and any other answer seems quite counterintuitive, we will consider other responses from contemporary religious thinkers in the next chapter, since these responses represent an important position in contemporary theological discussion of God's action in the world, and so in the debate concerning theistic evolution. However, evolutionary biologists usually (and certainly atheistic naturalists) are quite adamant that if chance plays a role in a causal series then there cannot be any intended purpose to the series. Any apparent goal or purpose is a matter of chance, not design. They are more ambiguous on the more general question concerning whether the universe in itself is deterministic or indeterministic, as we have noted. We need to bring out several further key points.

First, we can see from these concerns that it is a mistake to conflate randomness with unpredictability for the simple reason that a causal system could be fully deterministic, but so complex that it is unpredictable to us, yet not at all random because it might be moving toward a teleological goal, depending on its initial starting conditions and the laws that govern its operation. Randomness is a metaphysical claim made about how reality operates in itself; unpredictability refers to our knowledge of reality and whether we can discern how it will unfold.

Second, to say that some events occur randomly is not to be equated with the thesis that any event, therefore, is possible. It only means that

25. Other thinkers do not agree with this counterintuitive view, including Van Inwagen, who appears to think that if chance occurs, this means that no purpose can be intended; see *God, Knowledge, and Mystery*, 50–51.

one is claiming that certain particular events did not have to occur. An event is random if it is not the case that only one outcome is possible, given the earlier ingredients and the laws of science. Although it is a complicated matter, it does not follow that any series of events could result from chance processes present in the universe. This is a mistake made by Alexander.[26] It does not follow because chance is usually proposed as only a small feature of the universe, not a constant or dominant feature (if it exists at all!). Even those who rely a great deal on chance in their understanding of evolution, like Ayala, admit that the universe follows scientific law in a deterministic way *most of the time*.

Third, thinkers like Ayala never clearly bring out the relationship between determinism and chance and, as a result, do not properly appreciate the distinction between chance and randomness. What does Ayala mean when he says that the eye evolved in a random way? He means that the eye did not evolve with the teleological goal of being advantageous to the organism of which it is a part. Why does he conclude this? Crucially, because he thinks that at least some steps in its evolution *did not have to occur because of the presence of chance in nature* (in the way we have explained). Because these steps could have been different, we cannot say that the formation of the eye was aiming at a particular end (vision). To say that, each step would have to be necessary and would have to lead deterministically to the next step. However, Ayala also wishes to say that it is not true that the eye could have evolved *any way at all*, or following any path that we could conceive of, or, to put it more formally, that every possible evolutionary path forward is equally likely. No, he thinks that, *given the existing matter and ingredients and the laws of science*, some paths are much more likely (even in some cases predictable) than others. This is because, despite the presence of chance, nature operates in a law-like way most of the time.

However, this sweeping reliance on chance operating throughout the universe may present more of a problem for his claim that not all outcomes are equally likely than might first appear. It seems to be a problem for the narrow view of constrained or "directed" or "less random" evolution that Ayala, Dawkins, and others try to defend. In an attempt to downplay the notion of complexity coming out of a purely chance process, Dawkins proposed that natural selection be understood as "the

26. See Alexander, *Is There Purpose in Biology?*, 162. Alexander confuses the issue by suggesting that if all outcomes are not equally likely, then this means that none of them can be random. But this conclusion does not seem to follow.

non-random survival of random variants."[27] Ayala's example of the monkey typing, discussed in the previous chapter, was an attempt to make the same point. The difficulty they face is more obvious if we distinguish between considering, at the *beginning* of the process, what outcomes are likely to emerge and considering, in the *middle* of the process, the possible outcomes, when the evolutionary journey is well underway. It seems to me that all outcomes are equally likely (save for the constraints imposed by the nature of physical objects themselves and scientific law) at the beginning if both lines are governed by chance—the environmental line that leads somehow to the origin of life, and then the line leading to the development of organisms and species (sometimes referred to as selection and mutation). Later on, in the middle of the process, there are constraints on what can evolve, but this does not refute the argument that anything (within perhaps quite a wide range) might have evolved, since the beginning of the process determines every subsequent development absolutely. If you roll a boulder down a hill and allow that chance plays a role in where it lands, it is still true that the makeup of the boulder and of the terrain influences or restricts (or "directs") where it can go (as do the scientific laws—it cannot float upwards!). But suppose you say that the makeup of the boulder and also of the terrain are themselves due to chance, then it seems that you could have ended up with *any* type of boulder and terrain (or a completely different set-up) and so from this vantage point, myriad outcomes (if not all) appear to be likely. If ABCD constitute the initial raw stuff and place constraints so that only EFG can evolve, then the nature of EFG is indeed somewhat constrained, but what if ABCD itself occurs by chance? This line of thinking seems to go against their attempts to deny that evolution does not operate totally according to chance, and so the argument still holds that the evolution of complexity is impossible to believe if the whole process is governed by *pure chance*.

This is where the pressure is on evolutionary biologists to produce the evidence to support a view like Dawkins'. It is interesting to consider the question of how we *know* that the environment becomes increasingly hospitable toward changes in the organism, as Dawkins and Ayala claim. This claim invites worries about circularity—it looks like they might be saying that a complex species needs an increasingly hospitable

27. Dawkins, *Climbing Mount Improbable*, 79; in fact, this whole book is a series of superficial speculations suggesting how natural selection could produce complex organisms that appear to be designed. Not only are Dawkins' speculations far from convincing, but we can be sure that they are *not* the way things happened.

environment to survive, and rather than giving some specific evidence that would illustrate this claim, they insist superficially that if we have a complex species before us, it must therefore have developed in an increasingly hospitable environment! Upon discovering a complex species, we can infer that the environmental features that led to it *must* have been increasingly hospitable to it! To paper over this great difficulty, evolutionary biologists are often reduced to making up what I call "fictional stories" of how a species *might* have evolved, adjusting the story to include various hospitable environmental changes that *might* aid survival. However, these are not real, but fictional, cases—often called "just so" stories, as we noted earlier. The problem is even deeper, since providing direct evidence from real cases is impossible because the evidence that would be definitive is long gone. Yet, Ayala is confident: "In evolution, there is no entity or person who is selecting adaptive combinations"[28] (or who sets it up according to organizing principles). He also appears to acknowledge our point that any outcome is possible overall when he admits that the universe could revert to microbial prokaryotic cells—not just never evolve beyond these cells, but actually return to that early state, which seems to indicate that the process is governed by pure chance and that many outcomes are equally likely.[29] This claim is meant to emphasize his confidence in the apparently absolute random nature of evolution, a position that is very difficult to square with his view that the universe was designed by God.

Fourth, keeping these points in mind, it is obvious that biologists like Ayala's understanding of how evolution works *assumes* that the universe operates in an indeterministic way (similar to Gould). And we should note that when evolutionary biologists talk about indeterminism as a feature of the universe (which they almost never do but confine their thinking to the notions of chance and randomness which they do *not* consider in any depth or in relation to the larger issues), they are *not* referring to the biological changes that might occur at the level of quantum mechanics. They are talking instead about the behavior of macroscopic objects, such as geological features, wind, rain, heat, energy, water, trees and plants, animals and human beings, and atoms and molecules in DNA. (Indeed, while the study of this part of quantum mechanics, often called quantum biology, is ongoing, it is still very theoretical and

28. Ayala, *Darwin's Gift*, 71.

29. See Ayala, *Darwin's Gift*, 64–78.

controversial, especially in terms of the metaphysical question of how and even whether quantum effects have any influence over DNA, and therefore in the evolutionary process, more generally.)[30]

Fifth, this brings us to a crucial point concerning the relationship between determinism and evolution. If we accept determinism or even the lesser position that it is an open question, *we cannot then judge that an event that does not benefit an organism must therefore be random.* This claim is based on very controversial assumptions about how the universe operates, assumptions that have tremendous significance for the implications of evolution. It is not just because the question of determinism being an open question means that we do not know whether or not a quantum effect has any real influence on changes in DNA, nor that what appears to be a chance event in the causal process that leads to an outcome may not really be a chance event at all (if determinism is true).[31] This is not quite the right way to think about what is at stake. *What we need to say is that from the fact that an evolutionary change does not benefit an organism, it does not follow that the change was random.* We can only conclude that a change was random if it was not intended to occur. *And we can only conclude that if we assume an indeterministic universe.* However, I have argued at length that the universe operates deterministically. Evolutionary biologists may try to insist that they are not making an assumption that nature operates indeterministically, but are merely adopting a position that is based on empirical evidence, but I have argued that the evidence is against indeterminism (especially in ordinary physics and biology). Moreover, if they insist that they are nevertheless *interpreting* the evidence in the light of an assumption about real chance—that is, that it *looks like* evolution is operating by chance, so they will assume that it is—then this is a question-begging argument.

How can we conclude that a mutational and/or environmental change that does not benefit an organism may still be non-random? (For example, if an evolutionary change in the human eye led to the formation of a blind spot?) The reason is that there are two possibilities to consider, not one. We have to look at the issue in the right way in order to appreciate the point. In some cases, evolutionary changes benefit the organism (such as long necks in giraffes) and in some cases they do not (such as large antlers in the Irish elk). This is something we know empirically. So what conclusion

30. See the fascinating discussion in Mohseni et al., *Quantum Effects in Biology*.

31. See Ford, "What Is Chaos?"; also, Koperski, *Divine Action*, 39–69.

can we draw about whether or not the outcome is intended or occurs randomly? One conclusion is that both occurrences are governed by chance (in the sense that the causes of the evolutionary change could have been otherwise, and so the outcome might not have occurred). This is what evolutionary biologists mean when they say that changes occur without regard to whether they benefit an organism, and that there is no teleology. The second is that *neither* is governed by chance, and both must occur, given the prior causal chains. The second interpretation is the correct one *if determinism is true*. So if one goes with the first option, one is assuming an indeterministic universe.[32]

Now, assuming a deterministic universe, how should we think about advantageous and disadvantageous outcomes for organisms and species? There are two possibilities:

1. The universe was set up this way and both occurrences are intended by whoever set up the process (and the problematic ones—the ones that lead to a bad outcome—are part of the general problem of evil, a topic to which we will return in our last chapter).
2. The initial setup of the universe was itself a chance occurrence (one of our options from the previous chapter), but, since the universe operates in a deterministic fashion *after it begins*, then there are no further chance events in the universe.

This second reading is not believable for a number of reasons, the main one being *the complexity of the species that have emerged in our deterministically operating universe* (another is the anthropic or fine-tuning argument that shows order at the beginning of the universe). So the fact that mutations occur that are not always beneficial to an organism, that biological processes are less than optimum and disclose flaws and imperfections (as does nature generally) does not mean that there is no teleology in evolution (or in nature). The increasing complexity of the species overall is a (powerful) argument against the operation of chance. And the

32. Ayala's claim (noted earlier) that evolution, operating randomly, could eliminate all of our present species is true only if one assumes an indeterministic universe; if one assumes a deterministic universe, Ayala's scenario is *very unlikely* to occur (even though it could, since it is possible on *any* view of evolution—the key question is whether or not it is intended to happen). The existence of complex species counts strongly against Ayala's claim, which is why he struggles mightily in trying to explain how a random process led to such impressive complexity, especially (but not restricted to) *Homo sapiens*.

reply that this complexity only comes about randomly is to *assume* that there is no teleology, and so begs the very question at issue.

We must be careful not to define teleology in a narrow way to refer only to evolutionary developments that seem aimed at helping an organism adapt in a favorable way. This narrow understanding is not saying anything—specifically, it does not address the question of whether or not the process was *intended*. If you define teleology this way, then it becomes trivially true that disadvantageous mutations show no teleology. But this is vacuous because it does not establish anything about *purpose* in nature. Observing that a development does not benefit an organism is frequently used to argue that the evolutionary process overall is not directed or has no purpose. But this is question-begging because a process overall (*part* of which does not benefit an organism) could still be directed, especially since the organism *as a whole* has benefited favorably and is moving toward complexity. The whole tenor of modern evolutionary biology is to suggest that evolution has no overall purpose. But the evidence only shows at best that *some* mutations and environmental interactions do not lead to adaptation and favorable outcomes. That leaves the deeper questions of intention, direction, purpose, and teleology untouched.

Nevertheless, if one insists on the second possibility, one's case for the existence of a chance element in evolution is greatly diminished, because the chance operation of nature is now removed, and the presence of real chance is moved back to the beginning of the universe (and out of biology). This makes the question of how it all got started, chance or purpose, a live issue again, and the purpose answer is firmly back on the table. And no appeal can be made to the existence of chance in nature since, on this deterministic assumption, there is no chance in nature, so atheistic naturalism loses its (apparent) robustness and is considerably, perhaps fatally, weakened as a thesis about reality.

Overall, I want to emphasize that *one cannot say that the outcome of an evolutionary process is random if it leads to an undesirable outcome; for this assumes that the universe is indeterministic, and all indications are that this assumption is false.* Some biologists (like Ayala) may argue that intention is unlikely to be present if mutations and other evolutionary developments lead to bad outcomes. This, however, is not a scientific argument. Some may appeal then to various theological objections as to why bad outcomes could not be intended, or to the problem of evil in general as an argument against design/intention in the evolutionary process. These, however, are all different arguments (i.e., ones not based in science), which we will introduce below and discuss more fully in the next chapter.

Our sixth point is that it is a very significant claim indeed to conclude that the universe operates in a deterministic way because it raises the question of how it all began. I have suggested that the theistic answer is the most reasonable one. This is because of the complexity of the species that have come into being. The debate about whether this occurred randomly or by design is not a close one, as we have argued. And determinism adds to the strength of the theistic case because it eliminates the presence of chance in nature (or, according to some theologians, makes it negligible). The fact that the evolutionary process has produced complex species like us, with consciousness and self-awareness, and remarkable powers of reason, free will, and moral agency, beings who can not only understand the process that gave rise to them but who also have a fairly significant amount of control over it, strongly suggests that this is an intended and not a chance outcome. A species has emerged, in short, that can now direct, at least in part, its own evolutionary pathway. And to conclude that because (earlier) species could not do this, they must have come about by accident is simply to beg the question in favor of indeterminism and chance. Chance is possible, of course, but *it is not plausible.*

A kind of "front-loaded" form of theistic evolution is, therefore, viable. Indeed, it is the most viable option of those on the table, even if not proven. The conclusion of a deterministic universe, while perhaps not essential, does add considerably, I believe, to the viability of theistic evolution because it suggests that the universe is designed from the start to produce *Homo sapiens* as the end product of evolution. A serious worry about the existence of evil and suffering in the universe still remains, but this issue, while not easily dismissed and resisting easy or even any solution, does not diminish the status of theistic evolution as the most viable view (as I will argue in the final chapter). Theistic evolution is the best explanation for most of the features that we find in our universe, including *the truly remarkable phenomenon of complex, conscious, rational life.*

5

God's Action and Theories of Theistic Evolution

We have attempted to make the philosophical and scientific case for determinism and for the rejection of indeterminism and chance in the universe. Determinism is not only the most intelligible account of reality, both theoretically and practically, but the nature of our deterministic universe also suggests that it has been planned out by an intelligent mind. Any other conclusion seems to strain our credulity, for there is too much order and complexity to have come about by chance. Of course, the problem of evil still remains, but theism as a general view can cope with this, both logically and existentially, as it has done throughout its history. We will show in the last chapter how evolution not only does not make the problem of evil any worse (as some claim) but how it can be seen as alleviating it to some degree (though not because it operates by chance and thus removes God from responsibility). But the main point is that the rejection of chance operating in a deterministic universe opens the way for *a conclusion of design in nature*. This is the most reasonable response not only to the general order in the universe (the order that is the basis of the traditional argument from design and the fine-tuning argument) but also to the order and complexity that we find in the development of increasingly complex, sophisticated species, culminating in our own existence and nature. These realities further support the overall argument that theistic evolution is viable because one of the key claims that was intended to undermine it from the point of view of atheistic naturalism has been considerably weakened—i.e., that the presence of chance in evolution undercuts the notion of purpose and design in the universe.

Chance and determinism, as well as divine action and God's possible intervention in nature, are all related, so our next task is to consider these notions from the point of view of modern theology, which offers alternative theories of theistic evolution to mine. A theory of theistic evolution must not only be able to deal with the conflict between chance and purpose but must also provide a plausible conception of divine action so that God can achieve his purposes (including in evolution). We will see that one of the questions contemporary religious thinkers must address is whether and how divine action occurs, given that they wish to insist on the presence of chance in an indeterministically operating universe, a position that would seem not to leave much room for divine action, and certainly not for divine intention and purpose.

We are thinking about the way in which God acts to bring about his purposes. Although we deny that there is any chance operating, is it possible that God could allow a role for chance but still in some way direct or intend for specific outcomes to occur, including in evolution (one of our questions from chapter 3)? Contemporary theories that wish to allow some role for chance, but that also wish to uphold the view that God can act without intervening directly and without interrupting natural laws, struggle mightily with the problem of how to reconcile chance and purpose. I will argue that such views are not as viable as the position we are defending in this book (even if they can hold together logically). We will consider two such accounts in this chapter, in the work of theologian and scientist Arthur Peacocke and in the work of theologian John Haught. In the next chapter, we will consider several more views in the same vein that appeal more directly to the field of quantum mechanics.

We are defending the view that God set up the universe in a deterministic way, at least until the appearance of *Homo sapiens*. Cushing has suggested that the question of determinism/indeterminism is, in fact, the fundamental issue in relation to God's action in the world. He also argues that there is no decisive scientific evidence that settles the metaphysical question one way or the other, despite the fact that theologians often appeal to scientific justification (as we will see later).[1] Moreover,

1. See Cushing, "Determinism Versus Indeterminism."

we hold that the best way to understand God's action is that defended in traditional theism, that God not only can but does *intervene* directly in his creation (that is to say, God performs miracles from time to time). In doing so, say in responding to prayer, he suspends or overrides the laws of science. So I will defend the traditional position that God directly intervenes in nature by breaking or suspending scientific laws (the most reasonable view, given our deterministic outlook). We will explain over the next two chapters our position on how this works, as we contrast this more traditional view with various perspectives from contemporary theology, which either come close to denying that God intervenes in his creation, or propose a more elusive view that God does intervene but in a way that does not require suspending or "breaking" the laws of science.

Before elaborating on these fascinating but challenging issues, let us first consider, *in terms of specifics*, some of the ways in which God may act in nature. We will examine the question of what is often referred to as the "causal joint" by considering a range of cases and views. (This will help us to keep the discussion of various theological approaches, including our own, as concrete as possible.)

The "Causal Joint": How Divine Action Might Occur

Before we describe the general perspective of contemporary theology, it is helpful to consider in a commonsense manner the ways in which God might act in the world, what God's acting means, and how it should be understood. This will help us to bring out our own view more clearly, as well as sharply contrasting it with alternatives. Consideration of specific cases will help us to better understand the proposals coming from contemporary theology with regard to this question. Contemporary theological accounts that mostly reject direct divine intervention are quite weak with regard to the *specific* question of how God acts in the world; they often are more decisive and more confident in their descriptions of how God does not act rather than the ways in which he does act!

We should first consider the possible ways that divine intervention might occur. What exactly does God do to bring about change, say in answering a prayer? There are at least three distinct, though related, questions: (1) *Can* God intervene in his creation? (2) *Does* God intervene in his creation? (3) *In what way* does God intervene in his creation? Although most theistic philosophers and theologians answer yes with regard to 1 and

2, our main concern is with 3 (as it usually is for most theists who wish to avoid deism).

Let us elaborate first on what is often described as the "causal joint."[2] The causal joint refers to the specific place where God acts in the world, the connection or link between God and the action he brings about in the natural order. The important point is that *after* God acts, things occur differently than they would have if God had *not* acted in the way he did. The causal joint is, therefore, an essential notion because it means that God's action affects the course of events going forward, and so his action is significant and meaningful with regard to what occurs. It also highlights the fact that it would be quite vacuous to think that God could act in nature and yet for things to go on in future history as if he had not acted at all. It is not even clear that such a view is intelligible.

Second, we usually describe God's influence at the causal joint as a miracle. A miracle is sometimes defined as a suspension or interruption (or even as a "violation") by God in normally operating physical laws, in cases where we experience some unusual happening (such as a miraculous cure from illness), but it does not have to be understood this way. Any intervention by God in natural laws, even one that is unobserved or undetectable, may be described as a miracle. Miracles are not to be understood as eliminating the laws of science, but only as changing or overriding them in some rare circumstances, at a particular place and time.[3] This means that the essence or nature of a thing should not be understood to limit God's power over it (any limitation with regard to how God may act with regard to an entity comes from God, not from the entity).

Third, the notion of primary and secondary causality, an important distinction in Catholic philosophy since Thomas Aquinas, is relevant here, and we need to explain its place in the discussion.[4] Theists typically hold that God both creates and continuously sustains his creation. God is therefore the ultimate cause of reality, of all that exists or happens

2. This term was first introduced (pejoratively) by Farrar, *Faith and Speculation*, 65. Although we are primarily interested in the topic of divine action with regard to the role of chance, pioneering discussions in early modern theology can be found in Hebblewaite and Henderson, *Divine Action*; Wiles, *God's Action in the World*.

3. See Geivett and Habermas, *In Defense of Miracles*; also, Silva, "Thomas Aquinas and Some Neo-Thomists," where he discusses Aquinas' view that "a miracle is an act of God that goes besides or beyond the natural created causes" (2).

4. See Aquinas, *Summa Theologiae*, I.2.3; I.104.4; *Summa Contra Gentiles*, II.3–8; Dodds, *Unlocking Divine Action*, esp. ch, 1; International Theological Commission, *Communion and Stewardship*.

in the universe. Classical thinkers appealed to the distinction between primary and secondary causes as a way of elaborating this view. The idea is that God is *the primary cause* of the world, a world he created and that operates according to the laws of science, which he established. So God is the primary cause of all events. However, ordinary everyday events are governed by scientific laws and operate within the constraints of matter and energy (often referred to as *secondary causes*, which are studied in the various scientific disciplines). In addition, proponents hold that God does not normally intervene in the day-to-day operation of physical events (the secondary causes); secondary causes can act on their own; yet God can perform miracles whenever he wishes, and does so, especially by responding to prayer (it is sometimes overlooked that the creation of the universe would itself be a miracle).

Although classical theism rejects the deistic view that God set the universe in motion but never afterwards intervenes in creation, some think that the distinction between primary and second causes comes with a risk because the domain of secondary causality might suggest that the universe runs itself and so God is not needed (indeed, contemporary theologians often invoke the distinction in just this way to deny that God intervenes directly in nature, a misapplication of Aquinas, who did not use the distinction to deny that God does not act outside of natural causes). However, the distinction is useful because it helps to bring out and emphasize that when God acts in the world, he acts as primary cause, not as secondary cause (and so also avoiding occasionalism).[5] This is best understood by considering two questions: (1) Does any action by God take place in the chain of secondary causes? The answer to this question is Yes. (2) Is God's action itself a secondary cause? The answer is No—only God as primary cause can initiate an override of an otherwise lawful (deterministic) domain of secondary causes. In this way, God can temporarily "suspend" the laws—something creatures are unable to do. We might also ask if suggesting that God acts on secondary causes limits his transcendence or makes him a secondary cause? Obviously not, for the reasons given, the created order of secondary causes depends on him, but he is not himself a secondary cause and obviously does not depend

5. For further discussion, see Silva, "Thomas Aquinas and Some Neo-Thomists," 2–4; also Freddoso, "God's General Concurrence." Freddoso also draws attention to occasionalism, which many theists wish to avoid, and which he defines as the view that God alone causes the effects in nature; there is no secondary causation, and natural substances make no causal contribution (see 553).

upon secondary causes. Another key, but often overlooked, consequence of this distinction is that the universe and the events that occur in it only have limited power; in particular, a chain of causes cannot operate by itself in a completely mechanistic way. This is because the events must be sustained in existence by God initially. And so a complete explanation of how anything can come about requires recognition of the divine input. The discipline of science studies causal chains, but it leaves aside the question of the existence and sustaining of physical matter, so it cannot provide a complete explanation for reality from a theistic point of view. And so the notion that the universe can run by itself without divine assistance is rejected on this view, as is the idea that God might then be "absent" from his creation.

We can fill out some of these ideas by considering specific cases. Let us take a case of ordinary causation, say a pot of water boiling on the stove. According to the theistic view, God created the whole causal order of matter and energy in which an event of this kind can occur. He created the detailed environmental conditions, as well as the specific ingredients, sustains them in existence, and ordains that they behave in certain consistent ways, which we progressively identify as the laws of science. This is the initial setup, one might say. So God is the primary cause of the total scenario, including the water boiling. However, there is also the standpoint of the secondary causes. When we put water on the stove to boil, we don't understand God as *directly* intervening to cause the water to boil. The water boils because, given the divine sustaining, the ingredients, the laws, and so forth, this is how nature typically behaves. We might say that God is the *indirect* cause of the water boiling, perhaps, but he is not the *direct* cause. The stuff of the universe is governed for the most part by this kind of secondary causation, even though God is the primary cause, as we have noted. Theists also generally hold that God does not *normally* intervene in the process of secondary causation, although he is able to do so, but does so only *occasionally* (according to most theists). So, when we seek in ordinary experience the cause of the water boiling, from the point of view of the theist, we are looking for the secondary causes only, that is, for those causes that we study in science. We are not seeking the primary cause, nor are we asking for the human cause (except in some circumstances; for instance, if we ask "how did that door get open?" we often mean "who opened it?", which in this case could be described as seeking the main cause, rather than the subordinate, scientific answer).

What about the case of the turning of water into wine at the wedding of Cana, as described in the Gospel of John? We don't need to decide whether this account is accurate; we just need to think about how God, acting through Jesus, would bring about an event like this. In the Gospel account, Jesus told the servants to fill the large jars full of water. Soon, they realized the water had turned into wine. There are three things to consider here. The first two are obvious: that God, acting through Jesus, intervened in the chain of causation in the world and performed a miracle, and that God's intervention makes a difference—the world is different after his intervention than before it, in that there is now wine in the jars at Cana instead of water. But the third question is the most important: How exactly did God turn the water into wine? What did he actually do to bring this about? We might say that the simplest explanation is that by an act of will God changed the water immediately into wine; it is a kind of basic act, one that does not have a series of intermediate steps, an action that God brings about immediately, a direct act. God has the power to do this, of course, since he is the creator of all things, including the water, the wine, matter and energy, the environment, subatomic particles, the laws of science, etc. He is not himself subject to the laws that he created, nor in this case at least can he be reasonably said to act on the laws in any way, as a kind of invisible hand. (If one finds the invisible hand account attractive, then it is not hard to believe further that God could turn the water into wine by an act of will.) Apparent coincidences that occur in nature that we regard as miracles, such as surviving a car accident, may be understood in the same way, since (if the event really is a miracle) it would require some intervention by God in the laws of nature.

Consideration of two other types of events will help tease out the topic further. Suppose a huge dam has reached bursting point (based on ordinary scientific law and the natural behavior of objects), and is about to break open, leading to the flooding of a nearby town and significant loss of life. God intervenes and prevents the dam from bursting. God might bring about the outcome, of course, in any number of ways, but let us say that he causes a small tributary to break away from the bank of the river a mile downstream that causes enough run-off to reduce the pressure to safe levels. (St. Thomas Aquinas gives an example of God preventing fire from burning something.)[6] The tributary would not have formed if God had not intervened, and the dam would have burst. In this

6. See Silva, "Great Minds Think (Almost) Alike," 11.

example, the causal joint is where God brings about the small opening in the riverbank to allow a tributary to form. This would also be a direct cause by God introduced into the chain of secondary causation, not an indirect or mediated cause. Or suppose that God intervenes in the field of quantum mechanics and brings about a specific quantum event from a range of possible events, say he causes a particular radioactive particle to decay, which has such and such an effect (though even the notion of a "range" must be employed with caution, especially if it is used to restrict God's possible interventions). These are direct effects brought about by God as a result of his intervention in his creation. These cases prompt at least three more questions:

1. *Does this type of direct intervention by God occur, and can we perhaps provide more description of how it occurs in terms of specifics* (rather than simply noting that it is mysterious)? My answer, and that of most theists, to the first part is yes; and, for the second part, yes, we can offer some reasonable description of how these events might occur (and also of how they could not occur or of ways that God is not likely to use to bring them about!)
2. *Is this type of divine intervention something that we can detect when we examine the case*? Our answer is usually no, but it might be possible in some scenarios, at least to some degree, especially in situations that people often describe as "miracles."
3. *Can we discern the purpose of the divine intervention*? My answer is that in those cases where we suspect divine intervention, we cannot *detect* the purpose, but we can (of course) reasonably speculate about what it might be.

Let us push a bit further on *how* God might be understood to act in such cases. Millard Erickson has suggested one way to think about how God acts (say in our example of the river overflowing the dam) is that God might invisibly act as a force on the river to bring about a new tributary (in the same way that human beings might) and so preventing the dam collapse.[7] The difference from human intervention is that in God's case the force is invisible, yet it works the same way otherwise, with God simply becoming another scientific force that acts on the forces already present. This is an interesting suggestion, but it does not seem to apply in cases like the wedding of Cana because there, water was turned into

7. See Erickson, *Christian Theology*, 406–10.

wine directly and immediately; there was no intermediate causal process that led up to the change, as there is in the dam case. Even in the latter two examples just mentioned, it is probably better to think of them as coming about due to *an act of will* by God, rather than God acting as an invisible hand. One reason to support this view is that God acting as an invisible hand seems to make God another actor in the causal chain, and although not out of the question, this does not seem in keeping with the nature of God. God would then seem to be constrained by the physical laws he created (a view that seems quite odd). Some thinkers also suggest that this approach brings God too much down to our level in a way that diminishes his status.

With regard to the question of whether there would be any trace of God acting to change the water into wine, or in causing a new tributary to prevent the dam collapse, we should distinguish between traces in the sense of scientifically detectable traces, and traces in a logical sense that might be inferred from the unlikely (or apparently miraculous) nature of the event. On the basis of empirical evidence, we have to conclude that there do not seem to be any scientifically detectable traces in these events. What might such traces look like? One possibility is that if we understand God's intervention as that of an invisible hand, we might be able to make a logical inference from the fact that "the tree could not have fallen without a considerable force behind it"; or from "the tributary shows evidence of the bank being torn away to allow it to flow," to the fact that these cases suggest signs of divine intervention. But we would have to be quite careful to avoid the "God of the gaps" fallacy here. We have also noted that such evidence would seem to make God part of nature, or subject to his own laws, and so these are unlikely to be the kind of traces that would be convincing (though we must acknowledge as theists that God could intervene in nature in just this way if he so chooses). Perhaps we should put this point the right way round: for both scientific reasons and theological reasons, if we see a tree that could only fall if it had considerable force behind it, we would normally conclude that there is a scientific, physical cause responsible.

Some thinkers suggest that there might be a more sophisticated way of detecting God's intervention in a natural event. They appeal to the law of the conservation of energy, according to which the total amount of energy in a closed system remains constant over time. Their argument then is that we should be able to detect an increase in energy when water is turned into wine (if a force was involved from outside the system or an injection of physical energy occurred) or when a tributary is formed by deliberate intervention. They then argue (along with their atheistic counterparts) that, since no increase in energy is detected in this way, then God did not intervene, and all events likely to be interpreted as miracles are purely natural.[8] I don't find this argument convincing for two reasons. First, if God acts on natural causes by means of an act of will, then he will not leave any residual scientific traces to show that the laws of physics have been disturbed; he will turn water into wine and leave it at that (meaning that everything else that is measurable will be the same as if we had wine to begin with!). He will make the necessary adjustments to preserve lawfulness post-intervention. Even if God did act as an invisible hand, he could do so whilst keeping all of the other laws constant and leaving no scientific evidence of his intervention, For example, if God caused a tributary to form, he would likely adjust all of the other relevant laws, as well as the ingredients that exhibit the laws, so that they would hold in the usual manner immediately after his intervention. In this way, there would be very little trace of his intervention except to note that a remarkable, very unlikely event had occurred (but leaving it difficult to rule out the possibility, if not the probability, of it occurring naturally). God is more likely, I suggest, to do it this way, in order to maintain the universe as lawful and consistent. So any further effects he allows to occur because of his intervention would be negligible, since if you alter laws in one place it can affect happenings elsewhere that would not be part of his current purpose (in this sense, miracles are "isolated" events). (Referring back to our point about the closed system from the previous chapter: The universe is not closed on our view because God intervenes from time to time, but even though he intervenes, he maintains lawful consistency in how the universe behaves from the point immediately after his intervention.) In this way, nothing occurs in physical nature at least without God's willing it; he can alter laws and perform actions at any time; indeed, all parties to this discussion seem to accept that this is true of the nature of God in principle. Another way to express this point is that the objection based on the law of conservation of energy might apply if outside intervention into a physical system in our world came from an unknown physical source, or an extraterrestrial source, even a spiritual source, but it does not apply *if the intervention comes from God*!

8. See Koperski, *Divine Action*, 137–39, for a discussion of this topic.

The second reason is that I don't think God will make it completely obvious if he performs a miracle; in our appraisal of the event, there will always be room for doubt. Although these are hard issues to make a judgment about, I can see two arguments in support of this conclusion. One is that God may not wish to coerce people into belief (a familiar theological argument), which evidence over time from miracles might do. Second, God needs to keep the laws of science mostly consistent so that we trust them and believe them to be reliable. This is essential since we need their reliability in order to live, to progress, and indeed to realize all kinds of human values, such as compassion, relief of suffering through advances in medicine, and so on. The fact that we cannot be certain that a miraculous event occurred ensures that our general trust in the laws of science is not undermined, even though we as theists believe that God can and sometimes does override them (which is why many theists petition God in prayer for miraculous intervention in their earthly affairs).

These cases exploring the notion of the causal joint also raise the question of whether or how much we can allow *scientific constraint* into our thinking about the possibility of miracles. A central question would seem to be: When we are appraising an event as a possible miracle, how much does scientific thinking influence our judgment? This is not a question for the atheist, but it is for a theist, who both believes that God exists and that God acts in the world. It is hard to make much sense of petitionary prayer if one does not believe that God acts in response to at least some of our requests. The question is often whether a natural explanation for an unusual, apparently miraculous event is likely to be forthcoming, and reasonable people might disagree about this. But we don't want to rule out possible divine intervention by definition, as a matter of policy if we are theists, because of intimidation, say, by science, or because, for aesthetic reasons, we find direct divine intervention distasteful. So we must reject the view that no event can qualify as a miracle. Such a view begs the question at issue in favor of atheistic naturalism, and cannot be justified as a matter of policy. A second proposal, that science shows that no miracles can occur, must also be ruled out because science does not show this! There are many events that seem to be miraculous and so the topic would seem to be an open question. Of course, one could close it as a matter of policy, but this just brings us back to the first position.

This point also reminds us to be wary of a "God of the gaps" pitfall, an accusation often aimed at ID theorists. The critical point is that we should be careful in invoking God as an explanation for what seems like a "gap" in our current knowledge because it may lead to embarrassment later on if a natural explanation is discovered. This is an important caution, but it does not apply to most cases of alleged miracles because people are aware of the danger and don't invoke miracles lightly. Also, and second, there are some gaps that science will never plug, such as the question of the beginning of the universe or the origin of the laws of science. These too might be regarded as miracles on the theistic view, that is, events which occurred in the natural world that do not have (and in these cases cannot, in principle, have) a scientific explanation. This too is why philosophy and theology are always prior to science, because they address questions that science cannot answer. It is not so much a matter of science being unable to answer these questions; it is more a matter of not being in the conversation. So the question of miracles is still very much on the table. In what follows, we will consider a range of views, some of which suggest that God can act in ways other than the ones we have considered here, such as acting to bring about his various purposes, including human beings, without violating or breaking natural laws. Although counterintuitive when compared with the views we have just discussed, one way some thinkers suggest this argument could work is through quantum mechanics. We will examine some of these latter views in our next chapter. Let us now turn to consider the general approach to these questions in contemporary theology before examining the work of several influential contemporary theologians and scientists.

Contemporary Theology and Evolution

There is a group of contemporary theologians and theologian/scientists who are drawn to indeterminism because they believe it fits better with evolution and also because of what it seems to make possible theologically. While most scientists accept determinism and strive to hold to it, despite any problems, theologians regard its perceived problems for theology as obstacles to be overcome, and are therefore attracted to indeterminism. As Koperski insightfully puts it: "For most philosophers and theologians, determinism is an obstacle to be overcome. But physicists fight to keep it, despite anomalies."[9] Wildman acknowledges that for several modern theologians, "[indeterminism] is a necessary condition

9. Koperski, *Divine Action*, 2.

for non-interventionist special divine action."[10] It is not always easy to keep the theological objections distinct from the scientific objections in the work of contemporary theologians, so we will be sure to keep this important issue in mind in what follows. But this group of influential theologians is near unanimous in rejecting the traditional understanding of God's action in the world that we have been defending.

This distinctly modern approach is often referred to as "hands off" theology, with the idea being that theologians working from this perspective overwhelmingly favor the view that God *does not intervene directly* in his creation after setting it all in motion. That is to say, God creates and sustains the universe, but after that does not intervene in any *direct* way—he does not perform miracles, for example, such as parting the red sea, bringing about the virgin birth, turning water into wine, healing a person with a terminal disease, bringing good weather, preventing a natural disaster, or in responding to prayers. This view commits to indeterminism about reality, and accepts that there is a significant element of chance in nature, which can be seen operating particularly in the process of evolution, but also at the level of quantum mechanics. Perhaps subscribers to this approach logically do not need to accept indeterminism in order to reject divine intervention, yet they do so especially under the influence of evolution and quantum mechanics. As we have noted, we find different versions of this approach in a plethora of contemporary theologians, such as Arthur Peacocke, John Haught, Robert Russell, Nancey Murphy, Ted Peters, and others (though I don't wish to suggest that this is the dominant view in contemporary theology). We also find it in the work of several influential scientists working in the interface of theology and science, such as Collins, Miller, George Coyne, Polkinghorne, and others. A few are also attracted to chaos theory, even though (as we saw in chapter 4) there is consensus that it should be understood deterministically. Some thinkers still find it appealing because of the butterfly effect implication—that a very small change at one level might have large consequences in the macro world (and so they speculate that perhaps at the quantum level a small, undetectable, change by God could have significant macroscopic effects).[11] We will discuss the views of several of these thinkers over the next two chapters.

10. Wildman, "Divine Action Project," 40.

11. Polkinghorne thinks that the proper conclusion to draw from chaos theory is indeterminism, but, as Philip Clayton points out, this is an unpopular reading (see *God and Contemporary Science*, 195). Moreover, as we noted in the previous chapter,

Although contemporary religious thinkers have interesting and insightful things to say about many topics, they often find themselves in an odd position with regard to the relationship between evolution, science, and theology. This is in significant part because their scientific motivations seem odd and often seem opposed; on the one hand, they are steadfast in their belief that there are no exceptions to scientific law (which suggests determinism); however, on the other hand, they generally accept that chance is present in the universe in two areas, evolution and quantum mechanics. This apparent contradiction has to be explained and resolved, especially the evolution part. Here, they say that God allows for a certain amount of chance and so struggle with the question of purpose (some bite the bullet arguing that God does not know the final outcomes of evolution, such as Coyne and Haught; others argue for a mixture of law and chance, like Peacocke; still others obfuscate the question, such as Collins and Miller). With regard to quantum mechanics, thinkers following out this theory give it an indeterministic interpretation because they believe it can help with their negative stance on intervention (already adopted because of modern science). So, they commit to the interpretation that the quantum world operates indeterministically, that its results only reach, in principle, a level of statistical probability. This stance, they believe, enables them to say that God guides his creation without intervening directly (and so he would not be violating any scientific laws), and also with his intervention being undetectable in scientific investigation. Several *theological* reasons for why God does not directly intervene in his creation are often added, including. (1) the problem of evil; (2) God would likely not tinker later on with his creation; (3) God would not bring himself down to our level by intervening directly; (4) creating and sustaining laws, and then occasionally breaking them, seems contradictory. We will come back to all of these arguments in what follows.

This general approach can be traced back to earlier thinkers, such as Friedrich Schleiermacher, Rudolf Bultmann, Paul Tillich, and John Macquarrie, whose work emphasized the need to take due cognizance of the results of modern science (including evolution) as well as (later on) developments in quantum mechanics, in addition to commitment to a belief in chance. Bultmann is famous for saying that we cannot believe in miracles in an age of electric light and the radio, cannot avail ourselves of modern medicine, and yet believe in the New Testament

---

Wildman and Russell argue strongly that chaos theory must be understood in a deterministic way to make any sense.

world of miracles.[12] Bultmann's admonition has become more nuanced and sophisticated in recent thinkers who are struggling with several ideas that must be made consistent with each other, as they try to develop a coherent theory of divine action. They wrestle particularly with the question of the "causal joint"—that place or the precise way in which divine intervention (if it occurs) has an effect on nature, resulting in the future taking *a different path* than if God had not intervened. As already mentioned, they reject the view held by more traditional theologians that God directly intervenes in the laws of nature from time to time (thus "altering" or "breaking" or "suspending" or "intervening" or "acting providentially with regard to" the laws so as to bring about actual events that would not otherwise occur—our view in this book).

Alvin Plantinga has suggested that contemporary theologians are very intimidated by science, and I am inclined to think that there is some truth to this charge.[13] They seem keen to accept the dogmatic view of evolution that we see in Dawkins, Dennett, and Jerry Coyne (among others), almost without question. They do reject atheistic naturalism, of course, and thinkers such as Haught and Miller have done stellar work in offering trenchant criticism of that view. Yet, one finds little tendency to question any aspect of the current version of evolution, especially its claims about chance and randomness. As Peters and Hewlett have put it, contemporary theologians in the "hands off" group see acceptance of evolution as a ticket to get on the train of modernity.[14] Sometimes they seem overawed by the discipline of science (with justification, they would no doubt add). This seems to have the consequence not only that they must be seen to agree with and even welcome the theory of evolution (which is one thing) but also that they must acknowledge that it operates by chance, along with the view that we live in an indeterministic universe (which, as we have seen, are quite other things!). They often come close to speaking with disdain of traditional theological perspectives, describing them in what seem to be derogatory ways, as "perfectionist," "magical," and referring to any intervention by God in his creation as a "violation," or "interference," even as "intrusions" into or as "overpowering" what he had initially ordained! It should be clear that if one holds that God designed and intervenes occasionally in nature, it does not follow that one

12. See Bultmann, *New Testament Mythology*, 4.

13. See Plantinga, *Where the Conflict Really Lies*, 74.

14. See Peters and Hewlett, *Evolution from Creation to New Creation*, 120.

has a "perfectionist" view of God, and the term "magical" sounds like it is meant to disparage certain types of unsophisticated believers. Sometimes these terms seem more politically driven than theologically inspired.[15]

Such commitments generate a number of difficulties for the task of explaining how divine action occurs and how to reconcile it with purpose. One problem is how to get purpose out of an indeterministically operating universe, one that contains a significant range of chance occurrences. The second is how to avoid *deism*, the view that God creates the universe and then leaves it to its own devices, like a giant machine that runs itself. A "hands-off" approach would seem to favor a deistic view, but these thinkers wish to be *theists*, holding that God acts in the world (in some way) after creation in order to look after his creatures and fulfill his purposes (thus God's providence). So deism is not a serious option. Theism seems required in any reasonable Christian theology, and it is very consistent with religious tradition (which has always rejected deism as being incompatible with the biblical view, in particular). Yet, asserting that God acts in the world to bring about his purposes is difficult if you are a non-interventionist, who believes in a significant element of chance also. Science as practiced—and even in theory—can work with a deterministic view and even leans toward it, despite quantum mechanics, as we have seen. Yet, contemporary religious thinkers feel that they have to accept the current very dogmatically asserted interpretation of evolution as operating by chance, and so are drawn toward indeterminism about reality, toward quantum mechanics, and, to a lesser extent, toward chaos theory.

Koperski has helpfully classified theological views on divine action into three categories (instead of the usual two of interventionism and non-interventionism) to illustrate more clearly some of the positions being staked out.[16] The first category is the traditional approach, that of interventionists—those who think that God, by breaking or overriding or otherwise suspending the laws of science, intervenes in nature from time to time after creation, the view defended in this book (I am more concerned to defend determinism than interventionism, though, as illustrated earlier, a deterministic account *leaves room for God to intervene if he so wishes*). The second category, according to Koperski, is non-interventionism. Many modern theologians are in this camp. So how do they

15. See Clayton, *God and Contemporary Science*, 189. Haught uses the word "magical" to disparage the traditional understanding of biblical revelation; see Haught, *Mystery and Promise*, 34; also Van Till, "Fully Gifted Creation," 187–202.

16. See Koperski, *Divine Action*, 3–5.

avoid deism? By attempting to embrace (paradoxically) an almost metaphysical or at least scientifically robust view of chance, arguing that God intentionally builds chance into the universe in such a way that it may not frustrate the notion of purpose, but still plays some role in the outcomes. The presence of chance seems to make it unnecessary for God to intervene directly in his creation because events can take an unexpected path since reality is indeterministic. This view sounds extremely problematic on its face. It raises the question whether this kind of argument is logically intelligible and, even more important, even if it is intelligible whether it is *likely* to be correct (the question of *viability*).

Koperski goes on to divide this second category into two further groups to capture two distinct views, which helps to crystallize the issues more clearly, he argues (and I agree). The first category within the non-interventionist group we have already described—where a mixture of law and chance can be built in from the beginning, leading to teleological goals, at least in some general sense. But there is a second type of non-interventionism, a more radical stance—the view that God can act and bring about change in the universe *without breaking the laws of physics* (which Koperski calls "nonviolationism"). Those who hold this view usually appeal to quantum mechanics to try to show how this process might work. Although there is some overlap between these two categories, I agree with Koperski that these views are sufficiently different to keep them distinct. There are scientists and theologians and sometimes scientists/theologians in all three camps, though our focus here is on the latter two categories. The first category would include most of the traditional thinkers in classical theism, across the major religions, including Christianity, Judaism, and Islam. This list would include the scholars of traditional Christianity, both Protestant and Catholic, including all the fathers of the church, St. Augustine, St. Thomas Aquinas, Blaise Pascal, John Locke, William Paley, Samuel Clarke, Al Ghazali in the Islamic tradition, and Maimonides in the Jewish tradition, along with the majority of religious thinkers. It would also include almost all of the mass of ordinary religious people, who believe that God directly intervened in history, intervenes in nature and life at least occasionally, and regularly in response to our prayers. (The main beliefs of this group are cited in the Nicene Creed, affirmed by Christians for the last seventeen hundred years, including beliefs in the virgin birth, the resurrection of Jesus, and the resurrection of the dead.) The second two categories include the thinkers mentioned above whose work we are considering in this and the next chapter. (Koperski himself is a kind of nonviolationist, though his position is nuanced.)[17]

Why do contemporary theologians propose what seem on their face to be odd, even logically contradictory, views? We have already noted their *scientific* motivations: They accept an indeterministic universe, significantly governed by chance, especially evident in both evolution and quantum mechanics (though evolution is the primary driver of their approach). As Peters and Hewlett observe, the apparent absence of purpose in the origination of the species, including *Homo sapiens*, seems to give some theologians nightmares![18] In addition, they also think that modern science makes it very difficult to conceive of any suspension of the laws of science (and so they are drawn to accounts of divine action that affirm it but argue that it does not involve any suspension of the laws). (The apparent contradiction between their commitment to the view that the universe is indeterministic and significantly governed by chance and their claim that modern scientific evidence makes it difficult to believe in exceptions to scientific laws seems to have escaped them, an issue, as far as I can see, none of them face head on.)

These thinkers do have some serious theological concerns that force them to question the traditional view of God's design of the world and his role in nature. The first is that they see a problem with God creating a universe that operates according to scientific laws but then "interfering" with it later on to change its course. They often employ the word "interfere" in their critique, though its use is clearly controversial because, since it is God's creation, it is probably not appropriate to describe God's actions in his own creation as "interference." "Intervention" might be a more apt term, but "acting providentially to bring about a desired outcome" is probably better! However, some claim that to act providentially in this way might be "dangerously close to outright contradiction";[19] it seems also to imply that God did not design things adequately from the beginning (a view advanced by Leibniz). He suggested that God would not intervene later to bring about additional purposes, since he could have built these goals in from the beginning.[20] These objections would

17. He wishes to avoid saying that God "violates" the laws of nature; see Koperski, *Divine Action*, chs. 7 and 8.

18. See Peters and Hewlett, *Evolution from Creation to New Creation*, 27.

19. Wildman, "Divine Action Project," 38.

20. Leibniz thought that God would surely create fully the universe from the beginning according to his wishes (he used the word "perfectly," but this does not seem

not seem to apply when human beings come along since we could freely choose an outcome that might go against God's wishes (and so he might intervene to guide us in a different direction). Some, however, worry that this might be a violation of our free will, and so another reason for why God would not intervene.

A second and main reason why many believe that God does not intervene is the problem of evil. This is the argument that if God intervenes from time to time, as the traditional view says, why, as George Ellis puts it, does he not do so more frequently?[21] Several objections to the possibility of divine intervention fit into this category—they are versions of the argument that if God were intervening in his creation from time to time, he would not allow both moral and natural evil to occur, and therefore the most rational conclusion is that he does *not* intervene at all. This is why Ayala (as we have seen) and also Haught (as we will see later in this chapter) describe evolution as a "gift" from God to man—because they argue that it absolves God of responsibility for evil. This objection also encompasses their argument that if God had designed the species he would not have allowed flaws in the designs (such as an eye with a blind spot). This is a theological claim motivated by the general problem of evil. So the relationship between chance and non-intervention is also crucial; evil happenings can occur because we have a universe governed by chance and in which God does not intervene. An extension of this point is that if God intervenes in nature occasionally and is responsible for the direction of evolution, this seems to make the problem of evil worse, since God is then responsible for all of the cruelty and suffering that evolution reveals. Whether this is an overly optimistic interpretation of the meaning of evolution, and whether it does absolve God of being responsible for evil, along with the other issues raised above, are all topics we will consider as we examine the interesting views of these theologians in more detail. I will argue below that these contemporary theological proposals, while no doubt novel and interesting, even intriguing in some cases, are not as viable as the view I defend in this book. Among other

---

to follow). He disagreed with Newton, who thought that God needed occasionally to "reform" the movement of the planets (inferred from mathematical discrepancies). Leibniz thought God performed miracles not to meet the needs of nature but only the needs of grace; Newton's reply was that this position would make deism, and indeed, atheism, inevitable. For a perceptive overview of the dispute between them, see Lewis, "Flawed Theology and Philosophy"; also McDonald and Tro, "In Defense of Methodological Naturalism."

21. See Ellis, "Ordinary and Extraordinary Divine Action," 383.

critical points, I will further develop the argument that evolution does not operate by chance, that God intervening but not breaking any laws is hard to make sense out of, and that evolution understood as operating by chance does not help us with the problem of evil.

Chance and Purpose? Arthur Peacocke

Scientist and theologian Arthur Peacocke has developed an intriguing view that attempts to accommodate a mixture of chance and determinism in the universe. Unlike scientists George Coyne, Kenneth Miller, and several of the theologians that belong to what is known as the Divine Action Project,[22] who hold that God does not know in advance which species will emerge from the evolutionary process, Peacocke wishes to argue that, although we can allow for an element of chance, God still intends and still controls the final outcomes, at least to a significant extent. In short, he suggests it is logically possible that there could exist elements of chance in the universe, and yet that the final outcomes *must* (in some sense at least) occur. So he rejects the view that the universe is governed by chance to such an extent that God would not know any of the final outcomes. He also rejects the view that I favor—that the outcomes, up to the appearance at least of *Homo sapiens* (when free will enters the picture), are built into the universe by God and unfold in a deterministic way according to the laws of science, and that there is no chance involved in the process. Peacocke's view is clearly influenced by modern science, at least to the extent that he wishes to acknowledge some role for chance and randomness, but not to the extent that there is no *design* involved, nor that God gives up control so that his purposes are left in doubt.[23] No theist, Peacocke included, will accept that the universe is not designed. But a theist might accept that evolution is governed by a degree of chance. The important question then concerns *how much* chance, especially if one claims also that God guides evolution. This is the position that Peacocke adopts and tries to develop.

I find his overall position to be a little vague, and it is hard to be precise about what he thinks concerning the key concepts and their interrelationships. Peacocke acknowledges that a deterministic view of

22. See the extensive overview in Wildman, "Divine Action Project."

23. Peacocke's view is expressed over a number of works, including *God and the New Biology*; *Theology for a Scientific Age*; *Creation and the World of Science*.

science is still widely held today, but suggests that it seems to be more of an assumption than a defensible position. Although he seems to recognize a clear distinction between determinism and unpredictability, there are several instances where he conflates the two notions. This is especially true in his appeal to the world of the microscopic at the quantum level, where he makes much of the uncertainty principle of Heisenberg. He acknowledges that there is a range of interpretations of the uncertainty principle, but he stays with a mainstream consensus that seems to settle for some form of indeterminism. He seems convinced that the uncertainty aspect of quantum mechanics shows that there is an actual indeterminacy at the microscopic level in how nature operates. Although acknowledging the statistical nature of this indeterminacy, he nevertheless interprets it as a real indeterminacy and not just as a failure to achieve precise calculations. In this way, some elements of chance are introduced into the workings of nature.

To further develop this approach, Peacocke suggests that it is possible to distinguish several meanings of chance. One meaning refers to events that are unpredictable by us; they may be unpredictable in principle because there are too many complex variables, or only unpredictable *at present* because of gaps in our knowledge. He uses the example of flipping a coin to note that this is a perfectly orderly causal process, but because the variables are difficult for us to calculate with precision, the outcome remains unpredictable, yet it is still a deterministic process. This, however, would be a case of unpredictability, rather than chance. A second meaning originates from the application of the uncertainty principle, in the way we have already noted. There is also a third crucial meaning, which describes the fact that two or more otherwise unrelated causal chains can intersect by coincidence; the effects of these interactions can then be described as chance occurrences. This type of chance can change the causal history of the universe, including in the process of evolution, Peacocke claims. To illustrate, he refers to an example of a person being hit by a falling hammer as they leave a building.[24] This is not quite the best illustration, however, since it involves human free will; if we were to take a different example of ordinary causal chains interacting in the physical universe—say an asteroid hitting the earth, or a boulder falling on a dam, or a species being wiped out by a storm—it is not so

24. See Peacocke, *Creation and the World of Science*, 91.

clear that there is any chance involved in his third sense (as we argued in chapter 4).

Peacocke then applies this general view to evolution. Evolutionary development, he suggests, is affected by chance in two senses. First, changes in DNA are affected by the uncertainty of quantum-scale events. Second, the overall general effects on the organism depend on the accidental intersection of causal chains in its environment.[25] Because of these chance occurrences, we can say that the evolutionary process is random in the sense that it is not aimed at the well-being of the organism (the sense of random we identified in the previous chapter). So he is suggesting that there is a significant part of evolution that is governed by chance; evolutionary developments are unpredictable for this reason, even though he notes that there are still "trends and certain generalizable features"[26] that might in principle be predictable or at least intelligible after the fact.

Does this mean then that the outcomes of evolution are entirely governed by chance? Peacocke notes that these kinds of causal processes might seem like a confused jumble,[27] governed by large elements of chance, a prospect that has led many to conclude that any species that comes out of the evolutionary process does so by "pure chance."[28] Even though it might seem that the chance intersection of causal chains, along with the indeterminacy that affects DNA changes, must have the consequence that whatever comes out of the process could not therefore be planned or intended in any way, Peacocke resists this conclusion, a conclusion quite a number of theologians now seem to embrace (however reluctantly). He suggests that "what we call chance, whether at the micro- or the macro-level, operates within a law-like framework which constrains and delimits the possible outcomes." And adds: "If all were governed by rigid law, a repetitive and uncreative order would prevail; if chance alone ruled, no forms, patterns, or organizations would persist long enough for them to have any real identity or real existence."[29] The interplay of law and chance is creative, he argues.

Many effects emerge from within systems; these systems consist of their own structures, their environments and ingredients, the laws, all

25. See Peacocke, *Theology for a Scientific Age*, 57.

26. Peacocke, *Theology for a Scientific Age*, 67.

27. See Peacocke, *Theology for a Scientific Age*, 65.

28. See Monod, *Chance and Necessity*, 112.

29. Peacocke, *Theology for a Scientific Age*, 65.

interacting with each other, and *given the system*, there is a range of possibilities that can come about in creation. This range is limited, according to Peacocke, citing the work of several other theorists who have written on this issue (including J. Maynard Smith, Karl Popper, and H. A. Smith). Even though there is an element of chance involved, and we can't predict the outcomes of any system with complete accuracy, it is still the case that, because the system constrains what can emerge, the system can be said to have a goal, as it were. This is why we can broadly predict which effects might emerge from a system, even if we cannot describe the actual outcomes in terms of specifics. So even though there is chance involved, there are also system and law constraints; what we have then is a combination of chance and law, and so the outcome could still be *directed* by God. Peacocke holds that God writes potentialities into the universe from the beginning and then a combination of law and chance seems to bring some of them about. God does not intend the specific ones that actually come about, it seems, but he does intend something like them. For example, he might not have intended our *specific* form of intelligent life, but he did intend *some form* of intelligent life to arise. He also intended for free beings to emerge, an essential development if beings are to have a genuine relationship with God.[30] Peacocke also believes that this open-endedness permits pain and suffering because these are necessary for the development of life, consciousness, and freedom. God is present, however, because he shares in the world's sufferings: "*God suffers in, with, and under the creative process of the world* with their costly, open-ended unfolding in time."[31] So there is a flexibility and open-endedness in nature, which assists the emergence of free beings.

God's lack of specific knowledge and the limits on his control, Peacocke is keen to emphasize, are constraints that God places upon himself, for he could have created a universe that was thoroughly deterministic and in which he knows and intended all of the outcomes (or at least most of them, allowing room for human free will). So there is some sense in which Peacocke is suggesting that a creative process like the one he describes, in which he says that God takes risks, is preferable to the alternatives.[32] I am not sure if this means morally preferable; another

30. See Peacocke, *Theology for a Scientific Age*, 125–26; see also *Creation and the World of Science*, 197.

31. Peacocke, *Theology for a Scientific Age*, 126; see also *Creation and the World of Science*, 199–203.

32. See Peacocke, *Creation and the World of Science*, 197.

way to think about it is that it is scientifically preferable, which means more consonant with current scientific evidence that suggests some randomness in nature, perhaps (of which more in a moment). The dice, in Peacocke's account, are loaded in favor of life and increased complexity. In another place, he says that "chance appropriately defined is the means whereby the potentialities of the universe are actualized."[33] This general view, he believes, enriches our understanding of divine creation, and suggests that it is better than other, more deterministic and more providential views. His overall position has been described as a version of panentheism, which he understands as the view that "the world is regarded as being, in some sense, 'within' God, but the being of God is regarded as not exhausted by, or subsumed within, the world."[34]

To make such a view more plausible, Peacocke is attracted to a top-down view of causation. "The notion of causation," he writes, "when applied to systems, has usually been assumed to be 'bottom-up' causation—that is, the effect on the properties and behavior of the system of the properties and behaviors of its constituent unit. However, an influence of the state of the system as a whole on the behavior of its component units—a constraint exercised by the whole on its parts—has to be recognized."[35] This position offers an alternative to the usual bottom-up approach, dominant since Newton, which says that there is nothing in the whole (object, organism, organizational unit, etc.) that is not already present in the parts it is made up of; everything in the whole can be reduced to and ultimately explained, at least in principle, in terms of the parts. The top-down view argues that this approach is inadequate to explain all of the effects we observe in the processes and properties of chemistry, biology, psychology, and other areas. This account of causation is fascinating but controversial. According to proponents, the conventional reductionist approach is unable to explain everything we find at the higher level, such as the organization of biological systems, but also the nature of consciousness, the best example of what is often called an emergent view. There are many properties of consciousness (such as intentionality) that cannot be accounted for on a bottom-up approach,

33. Peacocke, *Creation and the World of Science*, 87.

34. Peacocke, *Intimations of Reality*, 79.

35. Peacocke, "God's Interaction with the World," 272; see also, Sperry, *Science and Moral Priority*, ch. 6; also Murphy, *Beyond Liberalism and Fundamentalism*, 149; see Polkinghorne's critique of this view of causation in his "Metaphysics of Divine Action," 151–52.

and indeed the failure of materialist views of consciousness, critics argue, considerably strengthens the view that the bottom-up approach, while no doubt a vital part of the structure of the universe, cannot be the whole story about the nature of causation.

With regard to how God might act in the world, the top-down view is quite vague, and often resorts to metaphors, with Peacocke's proposal that the universe may be part of God's body being one of the best known. As we noted, this approach from process philosophy does not help us with the key question of the causal joint—how God's action actually occurs, what God acts on, how he brings about a purposive move. If one is going to claim that there may be causal factors that come from the whole that can have some influence beyond what is contained in the constituent parts, but without violating scientific laws, one must give some account of what these causal influences are. Peacocke suggests that a "flow of information" from the whole can have an effect on the parts, without violating any natural laws, or without (apparently) there being any type of causal action from God to the world taking place. One can see that, intriguing though this suggestion is, it needs further details and elaboration before we can understand how it is supposed to work and function as a plausible view of divine action. The notion of top-down causation is an intriguing proposal, worthy of further study, but until we identify clear cases of causal influence and how they work, it must remain a research project for the future.

This general view, he believes, enables him to argue that God as immanent power can affect the outcomes in the world without any kind of external direct intervention—from "outside" as it were—being required. So God works *within* the natural processes, especially at the quantum level, and by means of the presence of some chance in creation. He employs the analogy of a musical composer to illustrate: Just as a composer plays and shapes a symphony by guiding it but not interfering in its internal coherence, so does God shape the universe.[36] In this way, God can influence events without breaking any scientific laws, a view sometimes described as "non-interventionist objective divine action," but he does not wish to say that God acts *directly* on quantum events, since this would be another form of direct intervention. His argument is more vague—that God can "influence" events at this level through statistical probability, but he does not really explain how or address the question of the causal joint in any

36. See Peacocke, *Theology for a Scientific Age*, 174–77.

detail, thereby leaving it hard to pin down his actual position on the crucial questions. He often seems to prefer metaphors, such as the composer and his symphony, to actual details.

Peacocke's arguments on chance and determinism are intriguing, and they have generated much discussion. Whether his vision of divine action is a plausible one, from a scientific and a philosophical (including a logical), point of view is another question. We might consider several interesting questions: (i) whether it is logically possible for God to create this kind of universe, one containing a mixture of law and chance; (ii) the strength of Peacocke's overall philosophical, theological, and scientific arguments; (iii) whether it is possible on a Peacocke view for God to still know the final outcomes in this universe; (iv) even if it is logically possible for God to create a universe that allows for chance occurrences, is it *believable* that God would set the universe up this way? In short, is Peacocke's overall view of theistic evolution viable?

Let us consider the first question as to whether it is logically possible for God to create a universe with a mixture of law and chance. One might think that God could do this given the assumption of omnipotence, and, as Peacocke suggests, he would only be freely placing limits on himself concerning what he could know about the future. But, as Thomas Aquinas noted, there are some things that even an omnipotent God cannot do—in this category he included the logically impossible; for example, God cannot create a square circle, erase the past, and so forth.[37] So perhaps this is one of those things? It seems that God might be able to set up a universe in which he did not know the future (assuming that his omniscience does extend to knowledge of the future), especially when we factor human free will into the situation (but that is a separate topic which I won't get into here). But could God set up a universe where he does not know the future because *he has designed* the universe so that it operates with a mixture of *law and chance*? And we might also add: where teleology—purposeful goals—is *retained*? This is quite a difficult question to make a judgment about.

As our discussion of determinism and indeterminism illustrated, it seems possible to have either chance or purpose, but not both. We might

37. See Aquinas, *Summa Theologiae*, I.25.3. Aquinas himself seems to have allowed for some degree of chance, but he did rule out that chance could give rise to purpose; see I.22.2; also, *Disputed Questions on Truth*, 5.2; also Silva, "Thomas Aquinas on Natural Contingency and Providence." St. Augustine denied the existence of real chance, see *Confessions*, III.6.

compare it with the notion of what are called random systems, such as in slot machines, random computer programs, random number generators, and so forth. None of these is truly random, however; such systems are only unpredictable, since there is a causal path in principle for every result that emerges; no action in the causal pathway that leads to an effect could have been otherwise than it, in fact, was. The symbols the slot machine rests at are fully determined by its internal mechanics and software (though unpredictable to us). Could God make a system similar to this but that was *truly random*, and not just one that was unpredictable? One in which there are some events that do not obey the usual scientific laws, at least some of the time, so that, occasionally, even in causal chains, the outcome is truly a matter of chance, *and might not have occurred (i.e., some other outcome could have occurred)*. Although I find this very counterintuitive, it is hard to rule it out, and it is difficult to say whether or not it is logically impossible, in St. Thomas's sense.[38]

Perhaps we could allow that God could design a sub-system that was truly random in the way Peacocke indicates, but that the larger system of which it is a part is not random, and the end result is intended by God. So perhaps God could design a sub-system within which he knows that intelligent beings will evolve but not the specific biological type (*Homo sapiens*). In this case, we would have to say that the intelligent beings are therefore intended by God, and, moreover, that the role chance plays is somewhat trivial and insignificant. So one might say that in this case, the set-up of the sub-system is designed, but not what it actually produces. Unless intelligent life emerges on purpose, by design, Peacocke's claim is weakened, as is his claim about nature having larger teleological goals. Now, is this kind of randomness possible for God? Perhaps it is, but it is hard to envisage how. We will allow that this kind of local chance within a sub-system *may* be possible, but it seems that Peacocke denies what we will call ultimate or overall chance. And it is this kind of overall chance that we must be concerned with if we wish to deny or affirm teleology in evolution. Peacocke seems to think that God does not know the future, not just because it might not be logically possible to do so, but also because the outcomes of the process of evolution are only probable but not definite. This is because there is a degree of chance that God has built

38. One philosopher who appears to rule it out is Cicero, who noted that God cannot know that which occurs by chance, because then it would not be genuine chance. He is defining chance as involving not knowing the outcome. This is one perspective on a complicated question! See Cicero, *De Divinatione*, II.vill.18.

into the process: "God's intention in creation may be seen as being fulfilled through a process whose detailed outcomes were not determined in advance but in which the emergence of complexity and eventually self-conscious beings was highly probable."[39] So, although God does influence what occurs, it seems he does not know the final outcomes on Peacocke's view.

It is also difficult to know whether we can describe this way of designing the universe as a type of "creativity" or as instigating a "process." Perhaps we could describe it this way if we allow that intelligent, sophisticated beings like us will definitely emerge, even though God does not know specifically what kind of beings, leaving the final outcomes to the serendipity of a mixture of law and chance. Yet, it seems that God's action could only be described as creative if he knew the outcomes that were coming; if they occur by chance, it is difficult to describe them as intended or as creative, unless we stretch the concept beyond its usual meaning (law and chance could be creative, perhaps, but would not lead to purpose). Another overall problem is that if God has preordained the general outcome, it is not clear that he would not then also have to logically preordain the steps to bring this outcome about, in which case no chance could be involved. It is a hard call as to whether or not this mixture of chance and law is logically possible, but Peacocke has raised a fascinating question. Even though it seems we can have either chance or purpose, but not both, it is hard to be definitive when we are thinking about the extent of God's power. One problem with Peacocke's account is that he gives us no clear examples illustrating the question of the causal joint, so that we could better grasp how he envisages that a happening governed by chance could yet lead to an intended outcome. (Perhaps this is because such examples are logically impossible to construct; any time an element of chance is introduced, the overall purpose is compromised.)

Peacocke thinks that if we adopt a deterministic view like the one I have argued for in this book, then we can only think of any further action by God in the world as an "intervention" in creation. He does not like this idea, but some of his reasons are vague. Why could we not suggest that God sets it all up in a deterministic way and then occasionally develops it to further his design plan? Let us not use words like "intervene," "meddle," "tinker,' "intrusion," or phrases like "finish it off," etc., because they seem to be based on an unearned prejudice toward this type of interaction. We

39. Peacocke, *Theology for a Scientific Age*, 119.

could just as easily look upon a deterministic/interventionist way of acting in a positive light. I don't see that it would diminish God or is not in line with other theological understanding. Many religious believers hold that God regularly intervenes in nature as a response to prayer, and occasionally also by means of miracles. The deterministic view would allow for this, for intervention, *at least in principle*. Peacocke would probably insist that God could logically ordain the *general* outcomes while retaining a degree of randomness that can bring about an unintended *specific* outcome, leading to what he refers to as a more "fertile" universe, and that this is better morally than a deterministic universe.

Even if it is logically possible for God to create a universe that allows for chance occurrences, is it *believable* that God would set the universe up this way? Is this a reasonable position to hold, in the light of some of our concerns? Is it a reasonable way to attempt to reconcile evolution and religion, the emergence of species through a process of chance, but that were in fact intended in some general way? There are some especially practical concerns: The first is that denying that God can intervene directly in nature does not absolve God of the problem of evil (we will return to this topic in the last chapter). A second concern is that God not having full control over the final outcomes is hard to accept from the point of view of common sense. Peacocke's attempt to rescue this view by arguing that there can be a mixture of law and chance is not a viable option, I contend, even if it may be a logical possibility. At the very least, he owes us an account of how chance and design could work together, at the microscopic or macroscopic level, to produce an *intended* outcome.

Evolution as Gift and Promise: John Haught

One thinker who vacillates on the question of whether or not God knows the final outcomes is John Haught, who has developed an interesting version of process theology from within the Catholic tradition. Haught is very good at painting a broad theological picture of how God and evolution can be reconciled, but keeps his exposition at such a level of abstraction that it is quite difficult to discern his views on the questions of importance. He favors a type of theological narrative approach often at the expense of analytical clarity. He too often avoids analysis of the topics that most concern our work here relating in particular to the notions of determinism, chance, causality, purpose, and God's specific intentions with regard to the process of evolution. One also detects a type of disdain for more traditional theological accounts, especially for recent creationist views and ID theory, but even with regard to traditional religious beliefs, including belief in the God of classical theism, in biblical miracles, and in miracles more generally as a result of God's action in the world. It is easier to identify those views he opposes than it is to be clear about his positive alternative! Over his many books, he seems to favor painting a type of poetic, theological-literary canvas more so than developing a rational theology. At this, he is very creative, but it also means that it is very hard to get a grip on his position. Haught has been a trenchant, effective critic of atheistic naturalism, and he is correct that theologians must do more to respond to the reality of evolution. But he goes much further, saying that "I cannot emphasize enough . . . what a gift evolution can be to our theology,"[40] no doubt an eyebrow-raising (if not entirely novel) claim from the point of view of traditional theists and indeed atheistic naturalists! One reason for this is that it forces us to rethink traditional theology, with Haught urging theologians to re-envision and possibly abandon key doctrines. (Of course, it is not so much evolution that is the gift for thinkers like Haught and Ayala as it is chance and indeterminism, since it is *these features* that make it possible to understand evolution as an undirected process.)

Writing under the influence of Alfred North Whitehead and process philosophy, and especially influenced by the views of Teilhard de Chardin (who described evolution as an "epic"),[41] Haught's view is best expressed by highlighting a number of main points. First, he suggests that the Abrahamic God's most characteristic attribute, as depicted in the Bible, is that of making and keeping promises, a perspective that fits in very well with the long history of the earth and the unfolding of evolution. Moreover, revelation, he boldly suggests, should not be regarded as "a magical intrusion of foreign information" or as a series of propositional unchanging truths that form a deposit to be preserved by the church.[42] Second, we should welcome evolution (since it comes from God) as an essential part of the cosmic story; it is good for theology "because it intensifies our sense that the universe is a drama to be read at different levels"; and indeed, we may need to recast all of theology in terms of evolution.[43]

40. Haught, *Responses to 101 Questions on God and Evolution*, 114.

41. See Haught, *Cosmic Vision*, 72.

42. Haught, *Mystery and Promise*, 34.

43. Haught, *Science and Faith*, 44.

The picture of God that emerges from this evolutionary theism is one who is infinite in Being, Truth, and Goodness and makes the universe intelligible.[44] Third, the theory of evolution allows us to think of life as drama rather than design; for a drama, however, Haught claims that three things are required: (1) There must be "contingency." (2) There must be a thread of continuity, that we can see (when we look backwards) from the vantage point of the end of the "story." (3) There must be "irreversible time" (the past is a "gift" that makes possible an "always new" future).[45] Theology, according to Haught, can interpret contingency, such as occurs in random genetic mutations, as a signal of nature's fundamental openness to new creation.[46] Fourth (without providing any analysis of how chance occurs or how it can be consistent with God's purposes), he suggests that "the existence of chance is what we should expect if the story of life is to be consistent with our trust in a caring and promising God who keeps the future open."[47] This claim suggests that God does not know the future—and specifically whether human beings will come out of the evolutionary process—though Haught does not address this question directly, as far as I can discern. But he is clear that the existence of chance in the process of evolution is real, and not apparent, and it is not to be confused with unpredictability.

Fifth, Haught holds that if we deny the existence of chance, then it follows that the world must be perfect from the beginning and that there is no room for serious change, what he describes as "novelty"; a perfectly designed product would lack spontaneity. He claims that "life requires the continual admittance of disruptive novelty."[48] Sixth, a God who sets the universe up this way is a loving God, and the old idea of God as designer is coercive (in some way), and this is morally undesirable. Seventh, he seems to suggest that there is a general advancement and direction to the universe as a whole, even though *Homo sapiens* is not heading toward any particular purpose/goal.[49] Eighth, with regard to the problem of evil, he is sympathetic to the view proposed by Ayala and others that interprets the evil evident in the process of evolution as a "gift" from God to us, and adds that, since God designed the world imperfectly, full

44. See Haught, *God and the New Atheism*, 50–52.
45. See Haught, *God After Darwin*, 100–102.
46. Polkinghorne holds a similar view; see *Faith of a Physicist*, 25–26, 75–87.
47. Haught, *Science and Faith*, 46.
48. Haught, *Deeper than Darwin*, 25.
49. Haught, *Responses to 101 Questions*, 110.

of chance, and so opening up the possibility for novelty, "we cannot be altogether surprised that imperfection, including the fact of pain, will be part of it."[50] A suffering God, he believes, is revealed through Christ's incarnation, passion, and crucifixion, where there is a forfeiting of control. This revelatory insight founds the belief that the God of the future enters into the myriad suffering struggles of all creatures as a companion.[51]

Although Haught's view is intriguing and described with a literary flair, it does not enlighten us very much with regard to the questions of this book, because he does not devote enough analysis to the key concepts of design, chance, contingency, determinism, and causation. Although he is correct that the concept of God's promise is a key theme in the Bible, it is a secondary theme when compared with the question of purpose. Surely an even more important truth is the fact that God exists, has a certain nature, and acts in the world, taking providential care of his people. The significance of these truths seems to me to be prior to the notion of the nature of God's promise and how he might fulfill it. I also don't see how it follows that if God designed the world, or we consider God as a designer, that he must then have designed the world in a perfect way and that there would be no room for novelty (a point already noted above). It is not clear in any case how a deterministic universe negates novelty. Moreover, novelty may still be introduced through free human action, assuming that we have a plausible account of how novelty is to be understood and accept Haught's argument that it is an essential part of God's purposes. Yet, we may still wonder if he is right that life requires continual "disruptive novelty," especially since for him this should be understood to include the existence of evil. A disconcerting feature of his approach is that he advances, often enthusiastically, claims such as these, but offers very little by way of argumentative support for them.

Haught does not lay out the alternatives well; the choice is not between his approach—a God of love, understood as persuasive rather than coercive, along with a universe of chance, spontaneity, novelty, and pain, vs. a God of coercive power, and a perfectly designed universe, with no really free options for man. He has glossed over the position that I believe we actually find ourselves in (the general theistic position): a God who, since he created and sustains the universe, must have unbelievable power, and who put some design and order into it as evident from nature (along

50. Haught, *God After Darwin*, 55.

51. See Haught, *Christianity and Science*, 63.

also, I contend, with the progressive nature of evolution), but also a universe that *may* (though as I argue does not) contain some chance, along with suffering and complex beings with higher rationality, free will, and moral agency. So if we accept that chance is present, the problem then is how to put chance and order together, and how to get purpose out of chance. With regard to this topic, the main concerns of traditional theology are how to reconcile chance with purpose, evil with divine goodness, and the specialness of human beings with a capricious evolutionary process.

Haught often appears to approach these problems from the point of view of "faith," meaning that he accepts as a starting point the truth of some form of theism, on trust perhaps, as it were, and then attempts to see how one can reconcile it with evolution. His goal does not seem to include demonstrating that God exists but to reinforce the plausibility of an already existing faith, a too facile approach, I suggest, to a complex matter.[52] The problem with this approach, which is also evident in the work of Alvin Plantinga, is that it sidesteps the question of whether evolution, or other evidence, can then count against the overall case for theism, and tempts one into not facing up squarely to the problems that evolution is thought to raise for religious belief, especially with regard to the notion of purpose. Haught's favoring of a type of "explanatory pluralism" also flows out of this general perspective. He argues that there are several levels of explanation for the goings on in the universe. For example, one can give the local cause as an explanation for an ordinary scientific event, an event, however, that also requires an ultimate explanation, which would be God. Although this is a very insightful point, his further claim that the explanations do not have to conflict with each other, while broadly true (this is what it means to say that science is not really in the conversation about ultimate meaning), runs into difficulties when you introduce an evolutionary process that operates, according to Haught, largely by chance. Chance then seems to conflict with purpose, a seemingly necessary feature of the overall explanation, God. At least, this is the way the majority thinking about this problem see it, so it is not an unusual or eccentric problem to raise, yet it is one that Haught's work fails to address adequately.

One of Haught's biggest errors is to suggest that the reason that many religious thinkers are suspicious of evolution is that it would leave

52. See Baron, "God Is Deeper Than Darwin," 655.

God with very little power over nature, or perhaps no power at all (and he implies in a disparaging way that their preference is for a "God of power," a belief he thinks somehow leads to many wrongs in society).[53] Whatever the truth of this latter claim, the main reasons I believe that people worry about contemporary articulations of evolution are the ones I noted in chapter 1, especially the problem of reconciling the chance nature of the process with a purpose for the human race, in addition to the question of whether the theory threatens the specialness of humanity. A further worry, perhaps for some, is the problem of evil. Haught's claim that traditional theists wish for a God of power is too facile. It would be more accurate to say that a God who created the universe for a purpose and who has certain attributes (such as considerable power) is difficult to reconcile with a process that operates largely by chance, and that seems to be, as many affirm, without purpose. These are the questions with which Haught, despite the grandiloquent, obfuscatory, enlightened contemporary picture he paints, never comes to terms.

53. See Haught, *God After Darwin*, 48–55.

6

Theology, Science, and Quantum Mechanics

Several contemporary thinkers have suggested that the strange world of quantum mechanics may provide an opportunity to develop interesting and perhaps fruitful theories addressing the question of how God acts in a world governed by scientific law, along with providing an opportunity to reconcile chance with divine purposes. These thinkers include scientist and theologian Robert Russell, theologian Nancey Murphy, scientist John Polkinghorne, and others, though they do not all approach the topic in the same way. We will discuss the work of some of these thinkers in this chapter. Thinkers like Russell are drawn in particular to an indeterministic interpretation of quantum mechanics because they sense in this unusual understanding of microscopic reality a possible way to develop a tentative set of suggestions for how God may act in the world *without* violating or breaking the laws of science, one of their main motivations in developing a theory of divine action. They will suggest that the uncertainties claimed for the quantum world provide an opportunity to conceive of how God may act, even though these actions would not be detectable by science.

Although such a view may seem contradictory on its face, some propose that it might work if we are able to develop a plausible account of how indeterminacy, chance, and purpose can fit together at the quantum level. Some theologians suggest that it is in the operation of chance (which they consider evident in nature) at the quantum level that we can locate God's action. This is where the "causal joint" would be located when considering what difference God's action makes for future events. One way of thinking about it is to suggest that God may *realize* at the quantum level those effects that we can only calculate to be *statistically probable*. Such action by God, then, it is claimed, would not break any scientific laws, since the effects that occur are *within* the calculated statistical range (which is also set up by God), yet the precise effect that occurs is not predictable by us, even in principle (if indeterminism about reality is true). Nor would it be detectable by us, since reality would still appear from an empirical point of view to obey scientific law (although this law is partly indeterminate)—that is, reality would appear to behave in the same way *as if God had not intervened* at the quantum level. This way of thinking also raises the question of purpose: Is it possible to affirm real purpose if reality is operating at the quantum level in an indeterministic way, and partly by chance? Can any sense be made out of the idea that God acts in that area where chance occurs, yet apparently with a purpose in mind? The challenges for this approach, as we will see, revolve around the notions of chance and purpose; in particular, how one obtains purpose from an indeterminate initial quantum state. For it would seem that if God, in some way, even if only indirectly, influences quantum events in one direction rather than another, then such events would not be indeterminate or operating by chance (this would seem to be true even if God only brings about a general purpose but not a specific path to that purpose—for example, a process leading to intelligent beings, but not the specific species of *Homo sapiens*). Moreover, God must be bringing about change for some reason—to fulfill some purpose, so this has to be factored into the account as well.

Quantum Mechanics and Divine Action: Robert Russell and Nancey Murphy

We will focus mostly on Russell's view, with a briefer discussion of Murphy's approach. It is fair to say that these theologians are in the same camp as Peacocke and Haught in the sense that they allow the results and status of contemporary science to drive the discussion (they would say justifiably so). They seek a space where science will permit theology to enter in a way that is consistent with science. They also wish to avoid the traditional view that God directly intervenes in his creation, holding that this approach is inconsistent with modern science; for Russell, such interventionism "threatens to minimize . . . credibility in a scientifically informed culture."[1] Although Russell agrees we must accept that God is not only a God who acts but also a God who acts for purpose—a God who providentially guides his creation—he notes that many think our choices with regard to divine action are restricted to two: We can either "affirm objective special providence at the cost of an interventionist and, in some extreme cases, an anti-scientific theology, or abandon objective special providence at the cost of a scientifically irrelevant . . . and tame theology."[2] We need, he suggests, a third option, and so he develops an approach he calls a "non-interventionist view of objective special providence." It is an unusual perspective in that he wishes to hold that God is active in nature (and so he rejects the notion that this is a "hands off" theology), but nevertheless the action of God occurs in such a way that it appears to us to be random and not to involve "breaking" or "intervening" in the laws of science. So these laws go on just as if God had not intervened, but as Russell notes, the principle of sufficient reason is also in play here. And so he argues that, from the point of view of this principle, sometimes God causes an event that would not have happened otherwise (even though the laws of science are not violated or broken), and at other occasions, the sufficient reason for the event can be obtained completely from natural causes.

To make such a view work, Russell appeals to quantum mechanics, which describes a strange, sometimes counterintuitive world, within which he thinks his unusual proposal may gain some traction. It must be stressed that although such a theological approach may appear strange, Russell and other thinkers pursuing this line are genuinely struggling with a difficult problem and tentatively suggesting possible, though odd, ways of dealing with it. They are trying to be responsible to science while at the same time endeavoring to show that the theistic God must be an active God, and so avoiding deism. They recognize that if one wants to say that God providentially guides history, then one must have a theory of divine action. This indeed is the foundation of the classical notion of providence; it is theistic, not deistic. They are also very mindful of the

1. Russell, "Special Providence and Genetic Mutation," 192.

2. Russell, "Special Providence and Genetic Mutation," 200.

problem of evil, and are trying to develop models of divine guidance and providence that may help with theories of theodicy as well, if possible. In this sense, we must admire their interesting proposals, even if, as I will discuss below, they face a number of very difficult problems that raise serious questions about their viability.

Russell accepts the conclusion also advanced by Peacocke that God acts with a mixture of law and chance in his creation. He argues that chance cannot be understood to mean "epistemic" chance, where it describes a gap in our (present) knowledge, but not real chance in nature (the type of chance discussed extensively in chapter 4). Correctly describing the epistemic understanding of chance as an "empty" notion, he observes that it reflects only our ignorance of causes and would not rule out regular divine intervention in nature. (However, as we will see, it is not clear how he avoids this general conclusion in any case.) In opposition to this view of chance, Russell argues that chance in nature and also in the process of evolution is "not a sign of epistemic ignorance but of ontological indeterminism."[3] While acknowledging that the debate is not settled, he believes there are strong arguments supporting Heisenberg's uncertainty principle. He refers to the various familiar examples noted in chapter 4 that are thought to suggest that reality itself might be indeterministic (e.g., covalent bonds in molecules, emission of photons by atomic electrons). When one of these possible effects becomes realized, we can discover the necessary but not the sufficient conditions that determine the precise outcome of the process.[4]

Indeterminism at the quantum level, however, as we saw in the previous chapter, does not mean that there is chaos at this level. As we have already mentioned, and as Russell notes, our research shows that effects at the quantum level still follow a statistical pattern, which means that those causal events we observe fall within a statistical range that is broadly predictable. While specific events are not fully predictable in principle and with complete accuracy, we know that there is a limited range of possibilities that can occur. Because we do not know which actual event will occur, Russell holds (along with many) that this enables us to conclude that *what occurs does so by chance*. This type of chance must be regarded as what I have described as *real chance* because the claim is that reality itself is indeterminate (according to this view of quantum mechanics)

3. Russell, "Special Providence and Genetic Mutation," 193.

4. See Russell, "Special Providence and Genetic Mutation," 203.

and not entirely predictable, even in principle (on a deterministic view, physical reality, when left to itself, is always predictable in principle). We noted in our earlier discussion of this position that even should this be true at the quantum level, reality at the macro level *is* entirely predictable (in principle) and follows scientific law. This may be a problem for Russell's view later on because, unlike some scientists, he claims that the quantum events can have a significant effect on macro events, including in the process of evolution, a general claim that is quite contentious.[5] According to Russell, then, God could act at this quantum level of chance and law, and the results would be "entirely consistent with the laws of science"; moreover, God's action would not involve any disruption in nature, since this is how nature operates "in itself."

Russell adds that God's action in this way is "mediated through the natural processes with which God works."[6] Mediated action by God is different from direct action in some way, though it is not always easy to keep these distinct. On the traditional view of God's action described at the beginning of the previous chapter, God's action in causing the tributary in the river, for example, would be a direct action because he makes the initial event happen (he opens a slight gap in the bank of the river); natural forces take over from there and continue to act in the usual consistent scientific way (and we will also stipulate that God knows how his intervention will turn out since it has a definite purpose). A mediated action, by contrast, would be one where God works out his purposes entirely through secondary causes, either in general or in specific cases, *without any direct intervention*. This would appear to require a deterministic universe, as we argued earlier (one in which God sets it up in an initial way and it then unfolds deterministically). Russell adds a twist to the idea of mediated causes by claiming that with the operation of chance at the quantum level, God is acting in a mediated way that is *undetectable* from the point of view of science. Yet, this means that God's action is *not* reducible to a natural process (otherwise God would not be acting). However, this position is to be distinguished from the more familiar view that even though when we examine nature scientifically, we cannot detect any guidance, God is still guiding things behind the scenes, as it were. This would be the classical (interventionist) position. By contrast, Russell

5. See Koperski, *Divine Action*, 42–59; Koperski suggests that "what goes on at the quantum level stays at the quantum level" (42). Russell acknowledges several remaining challenges for his view; see Russell, "What We've Learned from Quantum Mechanics."

6. Russell, "Special Providence and Genetic Mutation," 194.

is saying that God is causing, in an indirect way, actions at the quantum level that have effects in nature, and these effects would not occur if God did not cause them, even though from the point of view of science we see only random events.

According to Russell, these quantum events are not negligible when it comes to the macroscopic world, as many claim. They can have significant effects on the course of nature, including in the process of evolution. For instance, he says that quantum effects at the level of the genotype can then affect the phenotype, which can significantly affect the history and development of organisms and their environments. Moreover, many evolutionary changes come about by mutations in DNA, and these, he adds, can be brought into being by causes at the quantum level. The influence of God's action at the quantum level, therefore, can have a significant impact throughout the evolutionary process: "God has indeed created a universe in which God's special providence can come about in the long stretches of evolutionary biology without intervention."[7] God nudges the probabilities in specific directions, but does not micromanage every detail. Russell seems to hold that evolutionary patterns are not fixed in advance in this way, but that God's action and guidance ensures that certain events occur, such as life and consciousness.

Before probing Russell's intriguing view further, let us turn to the views of Murphy, whose ideas will help fill out further how God might act at the quantum level. Unlike some of these theologians, Murphy clearly distinguishes between unpredictability and indeterminacy, and she notes that some thinkers seem to think that if nature is unpredictable by us, it must be indeterminate in itself, a clearly logically fallacious move.[8] She agrees with Russell that any theory of divine action must be consistent with the Christian tradition of a God who acts with purpose, and must be able to account for Christian practices, such as petitionary prayer. Indeed, whatever view one holds concerning how God responds to petitionary prayers, or how petitionary prayer is to be understood, will usually involve taking a position or implying a position on divine action. She agrees with Russell that God cannot be understood as intervening directly in his creation. The problem with interventionist action is that it seems to neglect the qualitative difference between God's action and ordinary created causes, she claims. To make God a force that moves

7. Russell, "Special Providence and Genetic Mutation," 215.

8. See Murphy, "Divine Action in the Natural Order," 327.

physical objects seems to make God a part of the physical world.[9] This repeats the familiar but hardly convincing criticism that the interventionist position seems to make God a part of the system of physical forces.[10] Murphy also regards the existence of evil as a problem for the Russell/Peacocke view (which she refers to as an "immanentism") and disagrees with other thinkers such as Haught and Ayala, who ironically think the fact that God does not intervene at all is an advantage of their view with regard to dealing with the problem of evil.[11] This view motivates her to develop a distinctive position outlining how quantum mechanics can help us with these issues.

She argues that God acts in two ways in the universe: through the quantum world and through human intelligence and action. The apparently random events at the quantum level all involve (but are not exhausted by) specific, intentional acts of God, she claims. However, God's action at this level is limited by two factors: "First, God respects the integrity of the entities with which he cooperates—for instance, there are some things that God can do with an electron, and many other things that he *cannot* (e.g., make it have a rest-mass of a proton, or a positive charge)."[12] (Of course, these are limitations that God places upon himself.) Second, God still produces a world that to us seems orderly and law-like. Some interpretations of quantum mechanics suggest that quantum events occur randomly, but the question for the theologian, according to Murphy, is whether they are truly random or whether God brings them about. She notes, correctly, that the claim they occur randomly is counterintuitive and difficult to accept, and that a better option is to suggest that God brings about the events. We might put it by saying that God is a *hidden variable* in the causal process that operates at the quantum level.[13] This also preserves the logical principle of sufficient reason, which Russell's view struggled with, since he thought that chance is operating at this level of indeterminism, and yet that somehow God is responsible for the

9. See Murphy, *Beyond Liberalism and Fundamentalism*, 81.

10. For more on this critical point, see Alston, "Divine Action." According to Alston, theologians like Murphy "seem to think that if God shares any activity, status or category with creatures, that pulls Him down to their level. . . . Obviously there is a world of difference . . . between an infinite-source-of-all-being bringing about X, and you or me bringing about X" (53–54); see also Koperski, *Divine Action*, 150.

11. See Murphy, *Beyond Liberalism and Fundamentalism*, 81; also, Murphy, "Divine Action in the Natural Order," 331.

12. Murphy, "Divine Action in the Natural Order," 339–400.

13. See Murphy, "Divine Action in the Natural Order," 342.

events that occur. Murphy is more definite about what happens, and so avoids this conundrum.

Philip Clayton has expressed the general position well. He notes that quantum behavior is either (i) random or (ii) internally determined or externally determined by some physical entity or system or (iii) determined by God.[14] The first is hard to accept because of logic and the principle of sufficient reason; the latter two, he suggests, are unlikely according to science (or from a scientific point of view). Of these three possibilities, Murphy argues for the third option, divine determination of quantum events; in short, that God is the hidden variable. God can bring about teleological goals, such as the existence of human beings, by using purely scientific law. So we are not looking for something that science cannot explain as a place where God acts, because his action would not be a violation of natural laws. The criterion for an act of God on this model is not the absence of a scientific account. Murphy is also attracted to the view that there could be divine action taking place at the higher level of biology (top-down causation) that would not be detectable to science because the whole can be greater than the sum of the parts, and modern science is constricted by taking a bottom-up approach. She thinks that our changing view of science and the non-reductionist conception of the hierarchy of the sciences means that we no longer have clear scientific reasons for rejecting claims regarding special divine action. And there are many theological reasons to believe there is divine action, of course, since this is a *sine qua non* of most versions of theism.[15]

She argues that God's governance consists in activating or realizing one or another of the quantum entity's innate powers at particular instants, and these events are not possible without God's action.[16] Yet she seems to think that this does not involve intervention or violating laws, a problematic issue we will come back to in a moment. Murphy helpfully appeals to the notions of necessary and sufficient conditions here as a way to think about causation, suggesting that neither the natural nor the divine condition is assumed to be a sufficient condition;[17] neither God's action by itself nor the quantum effect by itself would bring about the final event; *both* would be required. God would be a necessary but not sufficient condition for every (post-creation) event, and the natural causes are

14. See Clayton, *God and Contemporary Science*, 217.

15. Murphy develops this view in *Beyond Liberalism and Fundamentalism*, 135–53.

16. See Murphy, "Divine Action in the Natural Order," 342.

17. See Murphy, "Divine Action in the Natural Order," 354.

insufficient by themselves to bring about the effects that actually occur.[18] (However, Murphy seems to overlook the fact that if we adhere to the doctrine of primary and secondary causation, it would seem that action on the part of God would always be both the necessary and the sufficient condition, this because he is responsible for the behavior of the natural causes as well, since he created them with their behavioral dispositions and sustains them in existence.) God's intentional act actualizes one of the possibilities inherent within the range of possible quantum effects. With regard to whether God acts at the quantum level regularly or only rarely, she believes that God is acting constantly in this way, and perhaps even *all* basic quantum effects are directed by God.[19] If God were acting only rarely, then he would be a mostly absent God, going against the Christian tradition, and she believes if he acted occasionally, he would be a competitor (though this latter point hardly follows, since God could decide for his purposes to intervene only occasionally in his creation). Because the efficient natural causes are insufficient at the quantum level to determine all outcomes, then God is not in competition with natural causes (it would seem, according to Murphy, that God would only be in competition with natural outcomes if they had a deterministic set of causes). For Murphy, God's action remains somewhat hidden at the quantum level because we need to have a reliable universe most of the time—a universe in which the laws operate mostly in a consistent way. This is for the simple reason that we can then make responsible choices with regard to human action and the way it has an impact on the world. However, this seems more an argument for God not interfering regularly than one for why God would not make his interventions more obvious, though perhaps there is little difference between the two.

What are we to make of Russell's and Murphy's views? We must acknowledge at the outset that they are making an interesting and some would say intriguing attempt to develop a modern view that takes full account of the pressures of empirical science and yet that tries to leave a space for God's action in the world. However, despite this, I believe these views suffer from several problems and also have a vagueness about them that makes it hard to discern the actual position of these authors on how God brings about causal change. I also find it very difficult to discern how Russell, in particular, can reconcile the role he gives to chance in the

18. See Murphy, "Divine Action in the Natural Order," 343.

19. See Murphy, "Divine Action in the Natural Order," 342.

operations of quantum mechanics with any place for purpose in God's action. This is where his view is rather unclear; it seems that one can either have chance or purpose, but not both.

However, an initial main objection to Russell is that it is very difficult to accept indeterminacy in nature at any level, including the quantum level, as explained in earlier chapters. Russell admits that there are several interpretations of quantum mechanics, and he refers to Polkinghorne's caution (and that of other scholars) that we should be wary of basing our theological views on any current scientific theory, given the possibility of revision in the future as we gain more knowledge. He acknowledges Bohm's more realist interpretation (referred to in chapter 4), recognizing that the debate is far from settled.[20] I have also pointed out that, faced with the probabilistic nature of the results at the quantum level, we have two alternatives. Since we cannot explain why a particular mechanical outcome occurs, but only its probability of occurring, as Russell notes, we must make what is basically a *metaphysical* decision either that when it occurs, it occurs by (real) chance (meaning that even given its prior causes, it did not have to occur); or second, we can conclude that the imprecision is only at the epistemic level. The second option means there are hidden, deterministic causes, even if, in principle, they cannot be discovered. Perhaps they are influenced by our observational and measurement techniques in a way that prevents exact predictions (a problem that still remains at the epistemic level). In this second interpretation, we reject the view that reality itself is indeterminate, not least on the grounds that an indeterminist view does not seem intelligible. This is why I very much doubt that any indeterministic interpretation can ever be established. In any case, it remains an open question, one reason why I think the arguments I am proposing in this book, which rely on determinism, are more plausible than Russell's.

In our discussion of Peacocke's view, we asked whether it is possible for God to allow any genuine chance in the universe. This question is relevant also for both Russell and Murphy, since both seem to indicate not so much that chance occurs at the quantum level, but, especially in the case of Russell, that *God is actually the cause* of the quantum events that we observe as occurring by chance. This is where Russell's view suffers from a vagueness that makes it hard to discern exactly how he thinks God acts at the level of quantum mechanics. It is unfortunate that he

20. See Russell, "Special Providence and Genetic Mutation," 218.

provides no detailed discussion of chance, scientific law, and purpose, nor of their interrelationships, nor of their role in God's action. This lack of specifics is quite typical of much of this theological literature; thinkers often avoid the key question concerning the causal joint and how it works in specific terms, glossing over the points of real tension in their views. Even though Russell's essay is sophisticated and wide-ranging, it leaves out more detailed attention to these topics, which seems to be vital for the intelligibility and plausibility of his proposals. For instance, he suggests that at the quantum level "faith sees God acting with nature to create the future" and "it is God fulfilling what nature offers, providentially bringing to be the future which God promises for all creation, acting specifically in all events, moment by moment."[21]

This action is neither a disruption of the natural processes nor a violation of the laws of physics, as we noted above. God is acting specifically here at the quantum level. This is an example of a remark that is, I suggest, too vague and prompts the question of how does God actually act at the quantum level? What is God doing with regard to the causal joint question? And also, what does Russell mean by "faith"—it sounds as if he is using it in the bad sense of "believing without evidence, or at least without regard to the evidence or rationality of the view" in question. Or is he using it only as a synonym for theology in this instance? There is a worry here because earlier he noted that there is nothing to be detected in nature when God acts that would show that God is acting; from the point of view of science, the happenings of nature are the same, so the theologian suggests that, despite this, God could be acting. Russell is drawn to this view—because he can claim that when we examine nature, its behavior does not reflect any intervention by God, yet (in some way) *God is influencing its course*. So in a way, he is trying to have his cake and also eat it because he can claim that God is not intervening in any direct way (that might in principle be detectable, or logically inferred), and yet God is still causally involved. God is bringing out those effects that are actually instantiated at the quantum level (within the limited range of possibilities that can occur)!

Yet, this view is problematic unless one argues that the course that nature follows is necessary (my deterministic view), for then one can look at the outcomes overall in nature as evidence for purpose. One can see the difficulty Russell's view faces: How does one get purpose out of quantum

21. Russell, "Special Providence and Genetic Mutation," 203.

mechanics operating according to chance at some indeterminate level? There are, in fact, two distinct problems: If we are talking about real chance it means that the effect did not have to occur and could have been otherwise; and second, there is no difference in what we observe one way or the other—meaning that what we observe seems to occur by chance, according to his view of quantum mechanics. Murphy does not seem quite as vulnerable to this objection because she suggests that God acts as a kind of hidden variable in causing the quantum effects that actually occur, so there does not seem to be much room for chance in her position, but it does seem as if God must act *directly* on her view and not indirectly (as she claims). Yet, her position is hard to discern with regard to chance, and she does not discuss the concept. But she seems to agree that chance plays some role in quantum mechanics, at least up until God's influence on the quantum effects as the hidden variable. Russell also claims God actualizes potential states at the quantum level,[22] but it is hard to see how God could do this *without intervening* in the causal chain.

In an attempt to bring out the problem with his view, we might lay it out this way:

1. At the quantum level, reality is indeterminate and governed by some elements of real chance; this means that a cause that brings about an effect could have been otherwise, and so the effect did not have to occur.
2. These effects can carry over and have significant impact in the macro world.
3. God acts at the quantum level by realizing particular effects from a statistical range of possibilities that could occur.
4. In doing so, he is neither intervening in nature nor violating physical laws because the effects that he actually realizes are always within the statistical range of possible effects.
5. God is directing reality, including the process of evolution, in a broad sense, even if not in a specific sense, so there is purpose in the universe.

The problem with this overall argument is that it seems very difficult to reconcile statement 4 with statement 3, and especially with statement 5. Surely if God acts to realize a particular possibility, he is acting directly

22. See Russell, "Special Providence and Genetic Mutation," 213.

or intervening in his creation; this is surely functionally equivalent to intervention in creation (no matter what one might claim in theory). It is not sufficient to say that the event which occurs does so within a small range of possible events, and therefore if God acts, he would not be intervening in natural laws. One might be able to say that God would not be directly intervening if the event that occurs does so by real chance, but if God *causes* it to occur—brings about one particular outcome rather than another—I confess I cannot see how this is not a clear case of divine intervention into the laws of nature. Russell might argue that the difference here is that God does not override or suspend any law, since the event was within the range of what could happen, and when we observe it, we do not see any laws being broken. But I don't think this is adequate. *The important question is whether God could influence a particular instance to be realized from a range of probabilities without determining the effect that actually occurs.* I don't see how this is possible. Either the event has to happen because of divine influence (even if God only intends a general rather than a specific pathway, in some cases), or it only *might* happen; if the latter, then purpose would seem elusive, and the outcomes must be seen as occurring randomly. If, on the other hand, the outcome is caused by God, and so *must* happen, we have direct intervention, the absence of chance, and an event aimed at a purpose. This means that there is no real chance operating at this level, but only apparent chance. After all, many miracles in the traditional sense (though not all), where we believe that God directly intervenes, involve situations where, when we examine them, we cannot detect anything outside of science, cannot see any *obvious* anomaly in how nature operates (the tributary case from the last chapter is one example). Peacocke has criticized Russell's view because he thinks that God must be seen as the direct cause at the quantum level; so does Paul Davies, who notes that the spirit, if not the letter, of the statistical laws of quantum mechanics are violated in Russell's proposal.[23]

Russell might try to insist that intervention can be understood in two ways: (1) The first is that God breaks a law. (2) But there is a second way where God does not break a law but realizes one of the natural effects that would happen anyway *within the laws*. My view is that this distinction cannot be sustained because if God realizes an effect, he has to make something happen that might not have happened otherwise, and this seems to involve the breaking of natural laws. Even if Russell were

23. See Davies, "Teleology Without Teleology," 153; also Peacocke, *Creation and the World of Science*, 95–96.

to note that in some instances at least the event that God brings about would have happened anyway because it is within the statistical range of probable events, this would only be a *coincidence* if God does not take direct steps to bring it about. If it just happens to come about anyway, then God is not the cause of it. Russell might press the point that there can be a difference between God suspending or breaking the laws of science and God influencing the laws in a subtle way. This might be true if it means that in some cases of miracles we can observe an anomaly and in some we cannot, but it does not work in the case of quantum effects because here God's influence is not empirically detectable, according to their view. If we press the distinction from the side of God, I must confess I don't see any difference between God directly breaking laws and subtly breaking them. In either case, surely he must intervene directly to bring about the effect/purpose? Mindful of their claims about the indeterministic nature of quantum mechanics, we might say that any input by God that realizes a specific outcome at the quantum level seems to require God's direct intervention in some way because natural laws operating according to themselves cannot bring about a specific outcome at this level (they can only do that if nature is *deterministic*).

Russell might argue in response to our objections that God could guide the range of quantum probabilities but not select the actual event that occurs. This would be to advance a view similar to Peacocke's—that a mixture of law and chance might be operating. Yet this view is vulnerable to the same problem facing Peacocke's view. The events that God actually realizes would seem to involve both his intervention and his manipulating the laws of physics. If neither of these things is true, as Russell asserts, it is very hard to see how God could realize a particular outcome (and so this seems to be God operating as a direct cause after all and not an effect coming about only due to indeterminism). He adds that God does not foreknow these results, but his view on this matter is too vague.[24] He struggles with this very serious problem throughout his argument. In support of his position, he quotes theologian Thomas Tracy's view that quantum mechanics allows us to think of special divine action as the "providential determination of otherwise undetermined events."[25] Tracy explains that "God's action remains hidden and takes the form of realizing one of several potentials in a quantum system," which sounds fine,

24. See Russell, "Special Providence and Genetic Mutation," 210–11.

25. Russell, "Special Providence and Genetic Mutation," 213.

but then he adds: "[and] not of manipulating subatomic particles as a quasi-physical force."[26] It is hard to avoid the conclusion that this sounds like double-talk. How does God realize a specific potential, that is, bring it into being, make it happen, etc., if he does not manipulate in some way the subatomic particles so that they behave in a way that *they would not have behaved in if he had not intervened*? It is not adequate to say that because the actual outcome was included in the range of possibilities, then—when it occurs—God brought it about *without intervening*. This seems to make no sense because if God really intends the event and brings it about, this removes the statistical *probability* in this case in favor of certainty (even if not epistemically for us as observers).

Murphy's position does not fall into these problems as much as Russell's, though she does seem vulnerable to the general difficulty that comes with saying that God can intervene in nature without violating natural laws. But she is clearly uneasy with indeterministic interpretations of quantum mechanics, holds that God realizes definite outcomes within a limited range of possibilities at the quantum level (because he respects the properties of objects and the consistent operation of scientific laws), and that God is responsible for all of the effects that are realized at the quantum level. Murphy is much less sure than Russell about how much the effects from the quantum world carry over into the macro world, and, as we have noted, this remains a contentious issue. Many believe that chance events at the quantum level are washed out at the macroscopic level.

But one of the key critical points facing her view concerns how much God can respect the integrity of his created entities to obey his scientific laws *while* (not if) bringing about effects in the world. This is, because, as with Russell's view as laid out above, it seems that when God realizes a *particular* possibility he would have to intervene at that point and cause something to happen that would not happen naturally, even if we allow that what happens naturally can happen in some cases "by (real) chance"—and so he would not be at that point respecting the usual way objects behave. If he were, then it seems that he could not cause a particular event. This is the dilemma facing this view. If God is to make a difference in the way things go at the quantum level, and cause an event that only *might* come about (because it is one within a range of possibilities), then it seems that God must in some way intervene in the laws of

26. Tracy, "Particular Providence and the God of the Gaps," 318.

science. And if he does this, he is violating or breaking or suspending the laws in the sense that they are not working in a purely natural way (even if they appear to be from our observational standpoint—because the outcome that occurs is one of those within the range of possibilities). We can accept Murphy's point that God would not alter an electron to allow it to have a positive charge, but even with this restriction he would still have to *make* it do something that it might not (from our point of view) have done itself, and in this sense then God must be intervening in natural laws (or, more accurately, that it *could not* do, given the present causal sequences of which it is a part). God, as Murphy acknowledges, may be the hidden variable, but surely is also the cause of the outcomes in an interventionist way? This is the general problem facing any theological view that appeals to the statistical nature of quantum mechanics in an attempt to argue that God can influence events without violating natural law, and I don't see any way around it. The problem would also carry over for Murphy in particular to the macroscopic world because presumably at this level, God would also respect scientific laws, and since this world operates more deterministically, there would be no (statistical) opportunity for God to intervene, and so God's action at that level would be ruled out completely. This is why these theologians put so much stock in the possibilities offered by the metaphysics of quantum mechanics, even though it is risky to tie one's view too closely to a controversial interpretation of a scientific theory that may be significantly revised in the future.

We should finally comment on Russell's claim that his view is not a version of the argument from design, and so is not part of a project of natural theology. He says that he is not trying to provide a rational case for a theistic view and its compatibility with evolution based on an analysis of the evidence in the natural world. He takes this line because he acknowledges that he cannot point to any evidence in the world that shows that it is different when God acts from the way it would be if God did not act. This would also be Murphy's position. As she notes, in their view, God can bring about purpose, such as the emergence of *Homo sapiens*, using purely scientific law and in a way that is not detectable (that is, that appears to occur by chance). Yet, this claim seems a bit counterintuitive, and perhaps they are being a bit disingenuous in claiming that they are not doing natural theology. It is true that they are not engaging, perhaps, in an argument from design or in an argument of the intelligent design form (where we have a gap in nature that we need God to explain). Yet, they are trying to show in general how God could act in a world of indeterminism and chance, and given a process of evolution governed by chance. Surely this is supposed to be a rational argument and would support the theistic case from the point of view of natural theology? Isn't Russell saying that God is acting at the quantum level and trying to give an account of how this can make sense?

This is where his vagueness about whether God knows the future of the evolutionary outcomes is important. If we could point to the fact that, aside from the general consistent behavior of nature as expressed in the laws of science, extraordinary, complex species with remarkable properties have emerged from the evolutionary process as evidence of design or of divine purpose, then we could argue that this fact reveals something of God's purposes. But it is not clear that Russell can or even wants to make such an argument; certainly, he is coy about it and vacillates on the question (as does Peacocke). He claims that God's action is mostly hidden from us, but that it may be discerned after the fact—which sounds like we can discern it in the way nature unfolds, but he does not elaborate.[27] Murphy acknowledges the worry that at the level of observed phenomena, scientific results would be consistent with deism and naturalism, but thinks this may be a strength because it keeps God's actions hidden from us, leaving a role for interpretation within the larger narrative of, and commitment to, the religious worldview.[28]

Perhaps Russell might agree that his work could be classified as natural theology in a very general sense. But his attitude is yet another example of the ambivalence of these theologians toward a philosophical approach and toward natural theology: On the one hand they want to build a rational case and to argue for some degree (evidence?) of purpose in nature, yet, on the other, they seem intimidated by the aura of modern science and its insistence on randomness in nature, and its lack of accommodation for divine intervention as a matter of policy. They are also very concerned not to be perceived to be on the side of ID, and traditional natural theology. In part, this attitude has led to these unusual theological speculations concerning how God might act in the world. That is why these current theories of theistic evolution, while certainly interesting and perhaps even valiant, are far from convincing!

27. See Russell, "Special Providence and Genetic Mutation," 197.

28. See Murphy, "Divine Action in the Natural Order," 352.

Design and Chance: Francis Collins and Karl Giberson

Several thinkers aim for an approach at a more popular level that attempts to reconcile religious belief, especially the notion of divine purpose, with evolution, understood as operating with a large degree of chance. The challenge is how to get purpose and chance to go together in a way that still makes reasonable sense out of the notion of divine providence, especially with regard to the coming into being and nature of *Homo sapiens*. Francis Collins, head of the Human Genome Project and former director of the National Institutes of Health, offers a less sophisticated, more nebulous version of Peacocke's view.[29] He has further developed his outlook in conjunction with physicist Karl Giberson. Preferring the term BioLogos to theistic evolution, Collins articulates his position by attempting to show how evolution and theistic belief are compatible. However, both in his own work and in the argument he develops with Giberson, he spends a disproportionate amount of time critiquing intelligent design theory and showing how theistic evolution is compatible with a metaphorical reading of the Book of Genesis. This is all to the neglect of a more detailed account of the vital concepts of determinism, indeterminism, and the role of chance in the universe, and in evolution in particular. Collins' concern with ID and traditional biblical creationism is a bit misplaced because his main worry should be how to reconcile design and chance in creation. After all, he is presenting this view as a serious alternative to intelligent design theory and to a literal reading of Genesis, so it behooves him to explain in a way that is plausible how God's providential action can be reconciled with chance in evolution. He does not, in short, give enough attention to the notion of viability.

Collins argues that human beings are intentionally created by God and are part of God's plan. However, he also appears to want to hold that evolution operates randomly in nature; also, more generally, he and Giberson argue that because of quantum mechanics, the universe operates randomly to some degree in itself. It is very notable that in Collins' account of evolution, and in the later work co-authored with Giberson, no time is devoted to a discussion of chance as one of the key facets of the theory. This omission is serious, since it leads, then, to a superficial treatment in their attempt to deflect criticism that the chance and random elements in evolution appear

29. See Collins, *Language of God*; also, Collins and Giberson, *Language of Science and Faith.*

incompatible with their claim that God designed or directs the process. Collins refers to Alister McGrath, who proposes the familiar argument that God could have worked out his creative plan by means of evolution, but does not discuss or explain whether this means, then, that all the outcomes of the process, including which species emerge, are directed by God, with no chance involved in the process. Collins addresses this objection about chance and randomness—that the species could be different than they in fact are—but replies that God could still be driving the outcomes.[30] So he appears to be affirming that there is, in fact, no (real) chance involved, but only what I have described as apparent chance; evolution would only appear, as Collins puts it, to be "a random and undirected process"; but would really be directed by God all along. From a vantage point outside of time as we know it, God would know which species might emerge (if the process were driven by chance), but Collins seems to think (erroneously) that it follows from this that God somehow *intends* these species to emerge (despite the presence of chance). He expresses the point this way: "In that context, evolution would appear to us to be driven by chance, but from God's perspective the outcome would be entirely specified. Thus, God could be completely and intimately involved in the creation of all species, while from our perspective, limited as it is by the tyranny of linear time, this would appear a random and undirected process."[31] This account, however, is simply too vague and appears to contradict his view that evolution operates with a large degree of (real) chance. If it operates with chance, how then is it directed intentionally by God? If it does not, then the outcomes of the theory are intended by God, and there is no chance. Which is it, in his view?

Collins and Giberson appear to accept that the kind of chance claimed for the theory of evolution is inconsistent with God's plan, incompatible with the view that there must be purpose in God's universe. They acknowledge the undue influence of Gould (as we have noted), correctly observing that he holds the most extreme view on the role of chance. Yet, they vacillate on the general question, and their position does not seem coherent (similar problems afflict the work of Kenneth Miller).[32] On the topic of determinism in the natural world, they rely

30. See Collins, *Language of God*, 204–5.

31. Collins, *Language of God*, 205.

32. Miller's view is developed in two main works, *Finding Darwin's God* (esp. chs. 7 and 8), and *Only a Theory* (esp. chs. 5 and 6). I think it is fair to characterize Miller's attempt to reconcile religion and evolution as one of the most vague among current

a lot on quantum mechanics, but frequently conflate indeterminism in nature itself with our inability to predict the outcomes in the quantum world. We have noted more than once that it is essential to keep these two concepts distinct. They are vague on whether the apparent indeterminate happenings at the quantum level actually affect reality at the macroscopic level, sometimes taking it as if they do and other times as if they do not (they acknowledge that the universe follows universal laws).[33]

These thinkers are also vague about what role, if any, quantum effects play in evolution itself. They adopt a concept from Polkinghorne that God has endowed the universe with a degree of freedom but do not explain clearly what this concept means or how it works. It seems to be a metaphor for indeterminism in the universe. In their refutation of ID theory, they are led into agreement with other contemporary theologians that after the universe is created and set up by God, he does not afterward intervene in any way in nature. Adopting this position of non-interventionism, they wish to hold, however, that the universe does not operate in a deterministic way. Perhaps mindful of the difficulties facing the views of Murphy and Russell, they appeal to Polkinghorne's concept of freedom. In the Collins/Giberson version, this seems to mean that physical matter in the universe, which normally obeys scientific law, has some degree of "freedom"; "many processes in nature exhibit a genuine unpredictability that looks, for all the world, like freedom. The behavior of electrons, both inside and outside atoms, is genuinely free in the sense that the behavior is *not* the result of outside influences or prior history."[34]

This should not be understood as similar to human freedom, which involves higher consciousness, where we can make deliberate, free choices that are outside science and not determined in the world of cause and effect. No, it seems to mean that there is some degree of *chance* in the way physical nature operates, so that not only is the course of nature not predictable by us, but in principle, there is no deterministic way it must go. However, they provide no discussion of the role of chance in the process, nor of the concepts of determinism, indeterminism, or even of quantum indeterminacy, to help us understand the notion of freedom and the role

theistic approaches, because of its overall failure to face up to the main questions; consequently, he fails even to show compatibility between God and evolution. His work is best known for its critique of intelligent design theory and for excellent overviews of the evidence for evolution (see the first four chapters of *Only a Theory*).

33. See Collins and Giberson, *Language of Science and Faith*, 117–19.

34. Collins and Giberson, *Language of Science and Faith*, 134–35.

it plays in physical causation. These thinkers suggest that while God is at all times involved in his creation, there is still a degree of freedom in creation. This sounds contradictory, and seems yet again to be a clear case of trying to have it both ways, a too superficial attempt to reconcile chance and purpose. Tracy has noted that if we allow that there is indeterministic chance in the universe, and even if we describe it as, in some way, "free," it is not the same thing, and is not yet *intentional* action.[35] Surely it remains a chance event until it is given direction, and Tracy suggests that even if we said that indeterministic chance is a necessary condition for God's activity, it is not, however, a sufficient condition, because God must still realize one of the possible alternatives so that it actually occurs. This leaves a problem for Collins, for it seems that we would be back to the same difficulty we identified above with Russell's argument, that if God *acts* to realize one of the possibilities (whether at the quantum level or not), this would require *direct* intervention, and the overriding of any chance happenings that may be lurking in the process. Freedom, in this account, appears to be a placeholder for an explanation they have not yet provided—how to get direction out of a chance process.

Collins and Giberson go on to claim that God "might influence the creation in subtle ways that are unrecognizable to scientific observation,"[36] which seems close to Russell's view, and sounds like God is directing things after all. They appear to be making the following two arguments, one about determinism and the other about purpose:

1. The universe is indeterministic, not deterministic.
2. This means that there is a degree of "freedom" or chance in its operation.
3. Yet, God could be *directing* creation in ways that science cannot detect.

Doesn't statement 3 contradict statement 1, since God is directing things after all, and so a process that appears to be indeterministic really is not (they insist that God is not interfering from outside, thereby "breaking the laws of nature")?[37]

Another way to express their argument is:

35. See Tracy, "Particular Providence and the God of the Gaps," 311.
36. Collins and Giberson, *Language of Science and Faith*, 119.
37. Collins and Giberson, *Language of Science and Faith*, 119.

1. Evolution shows that the species came about randomly.[38]
2. This is because the universe operates indeterministically, and so there is a degree of chance.
3. Yet, God is directing the process, so species actually came about by design.

Again, it sounds like 3 contradicts 1. This is especially true when we realize that "random" in statement 1 means random in the *ontological* sense, not in the epistemic sense. And 3 also contradicts 1 because it suggests that God does intervene, albeit in an undetectable way. The problem is that their view is not developed enough to meet the worries that evolution operating by chance poses for the question of purpose and direction in the universe. So this argument seems to face the same difficulties we identified above, especially in Russell's position.

It is important to emphasize again that their account has no discussion of the key concepts of chance, determinism, and indeterminism that would force them to address the questions and difficulties with the degree of detail that is required to advance our understanding. Near the end of their account, for instance, they address Gould's question of whether, if we ran the universe again, we would get the same result (a crucial question, as we pointed out earlier, involving indeterminism and determinism). In response, they appeal to Simon Conway Morris's theory of evolutionary convergence, where he argues that evolution operates in a somewhat deterministic fashion, so that if you ran it again, you would get much the same results as we have now.[39] Morris holds, among other things, that various structures have emerged independently several times (examples are eyes, wings, limbs), and this seems to suggest a directionality (and some kind of purpose or teleology) to the progress of evolution. (This view perhaps might be similar to saying that tornadoes have formed in

38. Collins in his book, *Language of God* (204–6), and Collins and Giberson, in *Language of Faith and Science* (114–20), continually fudge and obfuscate the question of the random nature of evolution, specifically, whether it is really random or just appears to be random. Science, they indicate, tells them the former, but they suggest the latter throughout their discussion, yet never face up to the question. Michael Ruse makes the same criticism of Alvin Plantinga—that he does not face up to the evidence from evolution, and so avoids addressing the central question (see Ruse, "How Not to Solve the Science/Religion Conflict").

39. Collins and Giberson, *Language of Science and Faith*, 202; see Morris, *Life's Solution*. Morris himself always seems coy about what can be reasonably inferred from the phenomenon of convergence about the question of design in evolution.

the same way many times—a phenomenon that suggests determinism; at the biological level, it seems to indicate the existence of some kind of biological-type laws, or regular biological processes that lead to predictable outcomes, such as in the process of photosynthesis or mitosis, as noted in chapter 4.) In invoking Conway Morris's intriguing research, Collins and Giberson are once again appealing to a deterministic-like theory to explain their view, which again runs contrary to their earlier position that the universe is indeterministic and operates with a significant element of chance. I suggest that this kind of superficial approach does not make the theory of theistic evolution that they proposing viable, because they gloss too superficially over the key difficulties.

7

The Viability of Theistic Evolution

It is time to bring together the separate threads of our argument to show that theistic evolution is viable; that it is, in fact, the most viable view among the alternatives! Of course, I hold that it is not only viable, but true, but viability is an acceptably modest thesis for now. It is all we need at the philosophical level to put our form of theistic evolution on the table as a plausible, serious option. The main thrust of our argument is that determinism is a reasonable position to hold about causal happenings, and so it follows that there is no chance operating in the universe. Hence, scientific research, as we saw, is always approached from a deterministic point of view (whatever theoretical ambivalence might accompany it). It follows from this general metaphysical position (with regard to this still-disputed topic) that the process of evolution, contrary to assumptions and superficial appearances, does not operate with any degree of chance. And so the outcomes that materialize from it *must* do so given the initial ingredients and the laws of physics (assuming no direct intervention by God along the way, or from human beings, since we can freely change its course, at least to some degree, in the way we have explained).

We will also show in this chapter that once one considers nature from this starting point and then examines closely the results of the evolutionary process, especially the emergence of sophisticated, complex beings like us, with higher-level consciousness, reason, free will, and moral agency, that the best, most reasonable explanation is that the universe was designed by an intelligent mind to bring about this outcome. This is an effective reply to the atheistic argument that the species came about by chance and there is no designer, for two reasons: First, there is, in fact, no chance operating in nature, as we have argued. Second, the argument that complex human beings came about by chance is not believable. There are two further problems facing the atheistic position. One is that there is very little direct evidence to support the claim that evolution produced complex species by chance, only a series of questionable inferences, and promissory notes, about what "must" have happened given some general facts and assumptions. We will come back to this point later, but it is important to remember that specific pathways and steps in macroevolution have not been established, even in a general outline. The second is the problem of how to explain the beginning of the universe. Naturalists talk as if physical reality operates on its own somehow and does not require an explanation. Yet, the distinction between primary and secondary causation (noted earlier) offers a quite plausible way to think about how the universe behaves and to answer at the same time the ultimate question of how it came into existence and how it got to be the way it is. This is an *intractable* problem for atheism because there *cannot* be, from a logical point of view, a naturalistic answer to this ultimate question. It is simply not believable to any fair-minded person that the universe we live in and have studied extensively in science could just come into existence out of nothing without a cause, nor that it could have existed forever.[1] (One can hardly say that there is no answer to the question.) Speculative (and non-scientific) suggestions involving a multiverse simply serve to underscore this point.[2] So even though we are not arguing for the existence of God in this book, this intractable problem for all naturalistic views strengthens the case for theistic evolution.

The absence of real chance is a significant part of my argument, and also, I contend, a very significant point about the universe. Even though

1. The literature on these questions, of course, is extensive. British philosopher Richard Swinburne has written a series of books on the rationality of religious belief (references in chapter 1). On the beginning of the universe, see Craig, *Kalam Cosmological Argument*; Craig and Smith, *Theism, Atheism, and Big Bang Cosmology*. On the fine-tuning of the universe, see Rees, *Just Six Numbers*.

2. This speculative theory has been criticized by a number of scientists for being unscientific, including Ellis, Roger Penrose, and Davies. For an excellent discussion, see Davies, *Goldilocks Enigma*, 151–90. Among other critical points, Davies argues that the hypothesis of the multiverse is not scientifically testable; it is really a metaphysical position, he believes, dressed up in the language of science, one that is often accepted on "faith."

it is still an open question, I believe that the conclusion that there is no chance present is the most defensible view, and the one that is most true to the way we practice science. It also makes the most sense metaphysically. Moreover, it is an effective reply to the atheistic naturalist—that there is no chance involved in the universe, either in physics or biology—since the atheist's whole case relies on an assumption of chance. This is a more reasonable response than to argue, as Plantinga suggests, that although mutations appear unguided, they really are not.[3] It is also a more convincing view than saying, as some theistic evolutionists favor (such as Collins), that although the atheistic naturalists are correct about evolution operating by chance, God still set the whole process in motion and still retains *control* in some way (because there is design in it *somehow*). Or that God exists, but has given up influence over significant elements of the natural world (because he has allowed the existence of chance events), and so does not know the final outcomes of his creation. I regard my view as more viable than these alternatives, which we have considered in the two previous chapters. Modern theological views are not viable because: (1) they fail to reconcile chance and purpose; (2) their solution to the problem of evil does not work; (3) they have not shown that there is anything problematic about God intervening in reality; (4) their claim that God can direct things without breaking physical laws is counterintuitive; and (5) their appeal to quantum mechanics, while intriguing, is too strained to be credible.

I contend that the most reasonable view for the Christian theist to adopt is that God set the universe up to operate deterministically for the most part (primary and secondary causality), but that for various reasons, he intervenes from time to time. This view makes the most sense out of occasional (apparent) miracles (which it is inconsistent for a theist to reject as a matter of policy) and out of biblical revelation. It also does not run into the considerable difficulties facing the view that God can influence causal events *without* direct intervention, or *without* suspending any scientific laws. It also makes the most sense out of the notion of God's providence. It is, as we have seen, very difficult to maintain any

3. See Plantinga's critique of the views of Dennett and Dawkins, *Where the Conflict Really Lies*, 3–49, 67, 272. A similar view can be found in Murray and Churchill who run predictability and chance together, particularly evident in their example of flipping a coin (14). Although they hint at a possible deterministic approach, their discussion of the random nature of evolution does not come to terms with the force of the objection because they conflate chance in the sense of being unpredictable (as does Plantinga), with chance understood as a real part of nature, which is what is claimed by their critics.

sense of purpose if one thinks that the universe operates indeterministically. In fact, if the universe operates indeterministically, then it would seem more necessary than ever for God to intervene to direct events that are otherwise at the mercy of chance along a certain teleological pathway, a position that our contemporary theological friends eventually fell into. If you bite the bullet, as Haught does, your view becomes less believable because God loses control of the final outcomes; you are then forced to paper over problems by claiming that from the end point we can see design or purpose that was not there from the beginning (which seems to be suggesting that outcomes were directed all along, and that the chance was apparent but not real).

What are some of the difficulties raised by both theists and atheists against the view we are defending? There are five, three of which recur in the work of the theologians we considered in the previous two chapters. The first is the denial of our thesis about chance. We have replied to this objection extensively in our discussion, and in any case at very least, it remains an open question. A second objection from recent theology is that the evidence from modern science is against the view that God directly intervenes in his creation. A third objection is that, if God intervenes, is he not bringing himself down to the lowly level of his creation (an objection raised by Murphy)? There are, fourth, other logical objections, such as why wouldn't God create a perfect universe to begin with (Leibniz' question in the previous chapter). Fifth, the problem of evil—if God can and does intervene, why does he not do so more often to prevent evil (as George Ellis has put it)? We replied to the second objection in chapter 5. We can reply to the third and fourth objections together by referring to some insightful points discussed by Ignacio Silva (writing from a Thomistic perspective). Surveying the views of various thinkers, including French philosopher Édouard Hugon, concerning why God might not create a perfect world right from the beginning, Silva notes several points. First, it would allow God to bring about some outcomes that surpass the nature of created things. Second, intervention by miracles is a sign to humans of God's divine perfection. Third, it also shows that nature is wholly subject to God. Fourth, God may be fulfilling in time what he has decided from eternity.[4] These reasons seem just as interesting and as persuasive as Murphy's point (earlier made by Leibniz). We will reply to the fifth objection above regarding the problem of evil later in this chapter.

4. See Silva, "Thomas Aquinas and Some Neo-Thomists," 9–11.

Overview of Seven Reasons Supporting Theistic Evolution

We can identify at least seven reasons that add to the cumulative case for the version of theistic evolution defended in this book. We will list them here and elaborate on them in the remainder of the chapter.

1. *Determinism and the absence of chance*: A deterministically operating universe has the consequence that evolution does not operate according to chance and also suggests design and purpose in the outcomes of the universe generally and in the evolutionary process in particular, culminating in *Homo sapiens*. This position has the advantage of affirming evolution, but promotes a more realistic approach to the understanding of chance than the views from contemporary theology we considered in previous chapters. It is also more plausible than the claim of atheistic naturalism that evolution is undirected.
2. *Most plausible account of divine action*: Another favorable consequence of the version of theistic evolution defended here is that it explains God's action in the world in a more convincing way. God is understood to intervene directly in his creation from time to time in a purposeful, non-haphazard way, in a way that is most consonant with religious practices, especially prayer, and in a way that satisfies our general intuitions about God and his creation.
3. *Better fit with biblical revelation taken as a whole*: Our position is a better fit with biblical revelation and with the understanding and interpretation of various doctrines in the Christian tradition. Although eschewing a literalist approach to Scripture and the position that every miraculous claim must be accepted as true, our approach does not deny the miraculous in favor of a purely metaphorical reading of main biblical themes, a position that serves only to undermine the text and almost fatally compromise the Christian theistic view of the world. Our approach also fits better with standard theological views relating to the incarnation, the resurrection, the atonement, providence, and the practice of prayer.
4. *Stronger, more plausible foundation for a response to the problem of evil*: Although it is a real problem and often appears both unsolvable and prohibitive to religious belief and commitment, our version of

theistic evolution offers a viable way to think about it, whereas the modern theological view does not.

5. *Differences in kind vs. differences in degree*: Our view also better supports, from a philosophical perspective, a point that is confirmed in our experience, the differences in kind that we observe between *Homo sapiens* and other species.
6. *Foundation for objective morality*: Any view that is based on God creating and intervening and guiding the universe for a particular purpose has an advantage when it comes to explaining the objectivity of morality. Modern views of theistic evolution always seem to be flirting with moral relativism because they inject so much chance into the development of *Homo sapiens*, and thus into culture and civilization.
7. *Meaning and purpose*: We must not leave out the crucial issue of meaning and purpose in life. I believe that our view is more viable than alternative views of theistic evolution, and certainly than atheistic naturalism, which offers us very little on this vital question.

General philosophical case for theism: We might just note a final point that would come under the heading of natural theology. Although the rationality of belief in God is not our main concern, it is possible, as we bring our main themes together, to build a rational, cumulative case for the existence of a God of purpose who is responsible for the universe and the appearance of *Homo sapiens*. This larger picture would include an argument for a designer, as well as the cosmological argument. It seems we are already halfway there with our arguments for the viability of theistic evolution.

Law, Chance, and Evolution

We have suggested that the deterministic approach is the most viable way to understand and respond to the question of chance in the universe. This is because determinism is the most reasonable view of how reality operates, and it is also the one most true to scientific practice. But one further considerable advantage is that it allows us to see that the universe is governed by purpose. We saw this in the various possibilities we identified in chapter 4, one of which we suggested is the most reasonable position—that God set reality up in a deterministic way, with no chance involved, and so his purposes work themselves out in the unfolding of the universe. This is at the same time an argument for design in the universe and an intelligence behind it, as we noted. One can also reason backwards from the deterministic nature of the universe and a consideration of the outcomes, including the arrival on the scene of enormously sophisticated beings like us, to the fact that the universe is most likely designed this way, and that this is further evidence of an intelligence behind it. This is a more reasonable view than the other possibilities—that the universe *just happened* to start off with its various ingredients and unfolded in a deterministic way that led to the present outcomes, or that the universe has a mixture of law and chance in some way and yet leads to designed, purposeful goals. I have already indicated that the atheistic position is not plausible, but we should say a little more about the other alternative that we have discussed extensively in the work of contemporary religious thinkers.

Our question here is not whether it is *possible* for God to create a universe with a mixture of law and chance, but whether it is *viable*. Is it likely that God did it this way? Is it a reasonable view to hold based on what we know both logically and from our study of the universe, including our work in science? We should distinguish also between the weaker view of law and chance that says that God allows some chance in his creation but still directs and knows the final outcomes (the view Peacocke and Murphy lean toward) and the more robust position that God allows for such a mixture of law and chance that he does not know the final outcomes and does not direct them (Haught and George Coyne). Though, as we have seen, some thinkers are vague on this matter and never come to grips with it (Collins and Giberson). Such views, I believe, may be logically possible, but are not viable because they run into too many logical and theological difficulties. We also acknowledge that God could create a universe that has the *appearance* of operating in part by chance, but where in reality there is no chance, the position that Plantinga appears to hold, but we reject that view also.

It seems logical to conclude that one can have chance or purpose in the way the universe operates, but not both. Yet Peacocke suggested that we could have both, as long as, in his view, we did not have too much chance, but Haught thinks we can have a good amount of chance, but must then adjust our understanding of God and theology to accommodate it, to a quite radical degree. I suggested in chapter 5 that although it *might* be logically possible for God to set up a universe with a mixture of law and chance, from our point of view, such a position seems logically dubious.

We can make a comparison to humans working on an experiment to produce a complex functioning organism. Would it be possible for us to design the organism but leave just one part of it to come about by chance, say one organ, an essential part in the functioning of the whole, and then let it integrate organically so that we would have a functioning, healthy organism? This seems not only practically impossible from our point of view but also logically impossible. Could God do something like this in the process of evolution—where a specific pathway is governed by real chance in some aspects yet leads to a teleological goal? Peacocke might say that we do accept that God can do things that seem logically impossible to us. Although this is correct, his view that an evolutionary process governed by chance could still be purposeful seems too far-fetched, possible perhaps but not probable. We could push our example further to say that when we examine the organism from an empirical point of view, all the parts exhibit organization, as does the integration of the parts, yet speculate that one of them came about by pure chance, whereas the others are designed. For human beings, this is frankly impossible. For God? It would not be enough to say that God could allow to occur by chance an event that is not crucial to the operation of the whole (in the way we might allow a haphazardly formed part to fit into the structure of the space shuttle, as long as it is not performing any essential function). For this would only be a case of apparent chance, since it would not affect the built-in purpose that the process is designed to achieve. Apparent chance does not count here, nor does unpredictability, nor the view that God is subtly directing things all along (even at the quantum level), and so forth, for all of these are simply the denial of real chance. And many of these suggestions are also saying that God subtly, but directly, intervenes, thereby contradicting the "hands off" theology that these theologians favor.

It is very difficult to countenance any view that says God does not have full control over the final outcomes—this is hard to accept from the point of view of the standard way to think about religion, and from the point of view of common sense. It appears to come very close to throwing the baby out with the bathwater, modifying our theology to fit a controversial metaphysical position about the nature of the universe. The more chance one allows into one's account of how physical reality behaves, the harder it is to conclude that a theistic view with a fairly robust teleology is viable. There comes a point for almost all of these theologians where they must retreat toward the conclusion that *overall* things are directed, and so this means that (in the end) chance plays a small, even negligible, role. Our view here is, I contend, more viable than this. Arguing from a position of determinism, and drawing the consequence that there is no chance, yet accepting the theory of evolution, it holds that the final outcomes are intended by God, who directly intervenes from time to time in his creation, especially in response to prayer (which we will come back to below). It also avoids a creationist approach to Scripture, while accepting the classical understanding of God's nature. Additionally, it eschews intelligent design theory because it is unnecessary for God to intervene directly at specific points to progress his creation since it follows a deterministic path from the beginning, leading to definite outcomes.

Biblical Revelation, Theological Doctrines, and Religious Practices

When it comes to the question of viability, one must consider seriously the question of how one's understanding of theistic evolution will treat traditional Christian themes and doctrines, such as God's relationship to the world, our relationship with God, the doctrines of atonement, incarnation, resurrection, providence, and other defining parts of the Christian worldview. It is reasonable to suggest that any theory of divine action should be consistent with key Christian themes, doctrines, and practices, such as worship, prayer, religious experience, sacrifice, spiritual development, eternal life, forgiveness, redemption, and others. Of course, we can only look at this matter in general here since a detailed account of Christian doctrines and religious experiences is outside the scope of our investigation, which is concerned in particular with how Christian theism is generally viable, given the acceptance of (our understanding of) the theory of evolution.

A central Christian teaching is that human beings can be in a relationship with God; indeed, that this is why human beings were created. This is perhaps the fundamental doctrine of Christianity, a *sine qua non* belief. A central biblical theme, it is expressed in many passages, including, for example: Genesis 1:26–27 (humanity is made in the image of God); Deuteronomy 6:5 (we are told to love God); Psalm 23 (God looks after us); Matthew 6:9 (Jesus teaches us how to pray); John 14:23 (love between humans and God); Acts 17:27–28 (our whole being is in God). Prayer is a central form of communication with God. If God exists as a being with a certain nature, it is quite easy to make sense out of prayer. We pray to God, who hears our prayers and responds, even if the response is not what we wish or always easy to discern. There are different forms of prayer, of course, such as prayers of thanksgiving, praise, intercession, confession, worship, and prayers of petition, where we make requests of God. From a logical point of view, all of these forms of prayer must involve a being who is capable of hearing our prayers and of responding to them in various ways. This view is most consistent with not only the biblical view of God but also with common sense, and with how prayer functions in our everyday practical religious life.

It is particularly instructive when thinking about the question of prayer to focus on petitionary prayers, prayers where we make requests of God, appeal to God, make demands of God, even beg in some cases for a favorable response. What are we requesting, demanding, appealing for, begging for, in these cases? Ultimately, we are asking *God to act in the world*. We are asking God to intervene in creation to bring about a state of affairs that very likely would not come about otherwise. This is what we are doing when we pray for recovery from an illness, for good weather, that we will pass our exams, for strength to endure a trial, for patience with a colleague, for a good crop of potatoes, for good health, for success in one's job, for peace in the world, for all the usual things people pray for every day. As just one example from millions, Pope Leo XIV recently prayed for increased vocations to the priesthood, for peace in the world, and for all mothers on Mother's Day in May 2025 (just a few days after his election).

We have already explained in chapter 5 the way in which it is best to understand the question of the "causal joint"—the way in which and the point at which God acts in order to effect change in the world. The most viable way to think about this is that God acts in a direct way in his creation from time to time by intervening, breaking or suspending or overriding usual scientific laws, thereby acting providentially to bring about effects for his creatures. So, if God acts, for instance, so that a patient recovers from an illness, or to bring about a good crop of potatoes, these are cases of direct intervention. We do not need to know whether God performs any miracles of this sort; we only have to affirm that if he does so, this is the way he brings about change—by direct intervention. (This is also the best way to understand the resurrection.) We have seen that, because modern theologians don't want to say that God intervenes directly, they are forced into very counterintuitive positions involving somewhat tortured accounts of how God acts through quantum mechanics; to be consistent, they also have to deny miracles (such as the resurrection) and propose alternative metaphorical readings of the biblical doctrines. If they retain such miracles, and sometimes they do, they risk undermining the position on divine action that they are so eagerly defending.

The truth of this point is even more evident when we consider prayers of petition that are designed to alter human behavior, including our own, for not all requests are about asking God to intervene in the course of natural events. What about cases where we ask God for peace in the world, or for patience, or to help us get a job? I believe that these cases are best understood by the theological doctrine of grace, where God influences human beings, but in a way similar to the way in which human beings influence each other, without any threat to or denial of our free will. Although there are different understandings of grace in different traditions, and it can be a complicated notion, it is relevant here because, with regard to human behavior, grace is freely bestowed by a loving God on individuals to help us live a good life, among other benefits. However, the individual must also cooperate with the gift of grace, must ask for help, and must make an effort in response. This is why I think it can help us understand how grace works if we compare it with the way human beings influence each other. Human beings influence each other by offering advice, arguing, attempting to persuade, cajoling, ridiculing, subjecting to (sometimes harsh) criticism, sarcasm, among other tactics! Often, these attempts provoke deeper reflection in the person they are aimed at. This is particularly true if an individual has sought out advice from someone they trust, but it can also occur, of course, when the advice is unsolicited. However, one must make some effort to listen to and reflect upon advice offered (in whatever form it is offered), and freely decide whether or not to act on it.

I think we can understand God influencing human behavior in the same way. So if we ask for help to pass our exams, we are asking God to inspire us to work hard enough to succeed; we are usually not asking that God alter the exam answers, since this would be unfair and immoral! Nor are we asking for God to override the free will of the examiners, though it might include God inspiring the examiner to grade exams fairly and responsibly, and so forth. So how does God do all of this in terms of the causal-joint question? I believe he puts thoughts and attitudes deliberately into our minds to help us reflect on things, either solicited or even unsolicited, but we are free to act on them, of course, so God is not causing us or forcing us to act. Our free will is still intact and at play here, just as it is when our fellow humans put thoughts into our minds. This is why public prayers can be important, as in the pope praying for peace in the world. In that case, we hope that in addition to God putting thoughts directly in the mind of those who are at war (which does not require that the prayer be public), the public nature of the prayer might also inspire those who are at war (or who have the power to influence the course of the war) to reflect further on what they are doing, and then, by means of God's grace, freely decide to pursue the path of peace. Although these matters are often complicated, haphazard perhaps in some way due to human folly, stupidity, stubbornness, selfishness, etc. (all actions and reactions indicative of our sinful ways), this is a sensible way to think about how prayer works.

We should not underestimate the difficulty facing the alternative theories we considered in the previous two chapters when it comes to the phenomenon of prayer. If one believes that God does not directly intervene in his creation, it is hard to make much sense of the notion of petitionary prayer, and perhaps even of requests for forgiveness, and so forth. And holding that God can intervene at the quantum level, operating through chance and probability, does not particularly help us. It might be possible to rescue this view if one were to argue that our prayers could influence God to causally influence reality at the micro level of quantum mechanics in order to bring about effects at the macro level, so that God is then altering the course of future events. However, as we noted above, this is really *direct* intervention under another name! This is why several modern theologians are pushed toward what we might call a metaphorical view of prayer, when they discuss the topic at all. This is especially true of Haught, who holds that the traditional understanding of prayer has been made redundant by science. He says that prayer is best interpreted as a trusting in the future that God is drawing us toward, that it should be regarded as an act of openness to the transcendent mystery of God.[5] This approach to prayer has also crept into recent philosophy of religion ever since the movement of Wittgensteinian fideism developed in the work of thinkers such as D. Z. Phillips and Don Cupitt. Phillips describes prayer in metaphorical terms as a type of trust and dependence on God (whom he also understands in a metaphorical way).[6] Indeed, Phillips and Cupitt go as far as to deny the objective, independent existence

5. See Haught, *New Cosmic Story*, 189–202.

6. See Phillips, *Concept of Prayer*.

of God. Haught, however, does not go this far, even though he denies the existence of miracles, rejects traditional understandings of biblical revelation, and treats prayer as mostly metaphorical. One is inevitably drawn toward metaphorical readings if one does not believe that God ever suspends or overrides the laws of physics.

Before we conclude this section, we need to comment briefly on another key Christian teaching—that concerning providence, since it is such a central doctrine for our work in this book. Although there are different accounts of providence in different denominations, the main idea is that there is an underlying purpose to the universe and especially human life that is intended by God. In both general providence (where God sustains and guides scientific law) and special providence (where God looks after individuals, sometimes by intervening through miracles), God's creation is under his general care and sovereignty. Providence is God's plan to bring creatures to their end, as Thomas Aquinas puts it.[7] The relationship of providence to free will is a topic for further discussion and has generated some puzzles and difficulties that are raised in the work of various thinkers (including Augustine). But one thing they all agree on is that certain outcomes are part of God's purposes in creating the universe, the main one being that humans are intended by God and, moreover, are at the top of the evolutionary tree by God's design and not by chance. For this reason, humans have dominion over nature. I suggest that these are bedrock beliefs of Christianity. Of course, that does not mean they are true, or that one could not propose other interpretations, including one that says that the species came about by chance, and that God may not even know which species will emerge in history, as some thinkers suggest. Could God have set it up this way? We discussed this at length earlier; our answer at best can only be *perhaps*, or *maybe*. Even so, I submit that it is not a *believable* view. It is quite inconsistent with Scripture, which, while perhaps not a fatal problem, is both worrying and counterintuitive for a Christian. The theory of evolution itself is not inconsistent with Scripture in the same way, so long as we understand that the process does not operate by chance. The claim that God himself does not know the final outcomes of the universe, or which species will come out of evolution, is very difficult to square with mainstream Christianity (to put it mildly), as we have already noted. Cardinal Schönborn has described this position, which he attributes to Fr. George Coyne, as

7. See Aquinas, *Summa Theologiae*, I.22–23.

"nonsense"![8] It is probably the one issue where most Christians would draw the line in any attempt to reconcile Christianity with modern evolutionary theory. The view I am proposing here safeguards these bedrock Christian teachings, while accepting evolution, and also God's direct intervention in the world. This view has the advantage of providing us with a very plausible understanding of the Christian practice of prayer. These points, I contend, strengthen the viability of this view considerably.

The Problem of Evil

The problem of evil has vexed religious people from the dawn of time. As we know only too well, the worry concerns why an all-good, all-powerful, and all-knowing God would allow evil into the world. With regard to our analysis, a particular worry concerns natural evil, evil such as earthquakes, disease, and natural disasters that cause suffering for human beings and animals. Such events are usually not the result of human free will and so are in a different category to moral evil—evil events caused by our free decisions that lead to suffering (such as murder, rape, and robbery). There is no need to document the existence of natural and moral evil, for it is part of the human condition; a consequence of it is much suffering, stress, and anger as we attempt to cope with it in ordinary experience. Man's inhumanity to man can perhaps be explained by the "free will defense"—the argument that the possibility of moral evil is the price paid for endowing human beings with real freedom (rather than "pretend" freedom), the highest gift that God can give to his creatures. Although some have found this explanation debatable, no such straightforward response seems available for natural evil. Moreover, many are inclined to think that the process of evolution makes the problem worse than it already is. This is because it reveals a process of predation, parasitism, and suffering at the heart of nature that has been going on for millions of years. This conflict, seemingly built into nature, has led to much suffering in the animal world; it appears that nature has been set up for species to prey/feed off each other in the process that Darwin's theory identified as "the survival of the fittest." This degree of suffering seems unnecessary to achieve any goal. Whatever one might claim about how evil might affect human beings in some positive way with regard to questions of truth,

8. See Schönborn, *Chance or Purpose?*, 169.

meaning, spiritual, and moral development, no such arguments can apply to animals, and so this type of evil appears pointless.

Of course, such considerations have led some to unbelief, but that is not our focus here. We are interested in whether modern theological responses to this problem are more viable than our more traditional response. The first point we must acknowledge is that while we don't need to dwell on the facts of evil in the world, there is no denying that the problem is a serious one. It cannot be downplayed, shoved aside, ignored, or lessened. We must confront it head-on and develop a response to it as theists. The second point is that even though the existence of evil does drive some to atheism, it usually does not have this consequence for the vast majority of religious believers. It may even be fair to say that there is no clear solution to it—the problem can only be managed and, for some, perhaps not even that. Yet, for most people, evil does not lead to unbelief, so the existential problem does not usually compromise the efficacy of the worldview of Christianity. While the fact of evil often leads to doubt, frustration, irritation, anger, defiance, belligerence, horror—all seemingly rational responses—it seldom leads to unbelief. While evil is very troubling and often puts a strain on religious commitment, the fact that it seldom leads to the abandonment of the religious view of the world shows clearly that the traditional theological response remains adequate for most people—that God must have a reason for evil, even if we cannot discern it in this life. Some, of course, go further and speculate about what the reason might be, but for most believers this general response functions as an acceptable, if not satisfying, rational response. Most believers are confident that the issue will be satisfactorily resolved in the next life (the ultimate triumph of good over evil). This is the standard way of thinking about the problem of evil in the history of Christian theism.

The views we have been considering in the last two chapters, however, reject this response. They are very much taken with George Ellis's question as to why God intervenes so seldom to prevent evil from occurring. Ellis's view reflects very well the way most believers think about evil in the world. Indeed, the absence of regular intervention by God is the source of their irritation, frustration, and anger at the presence of evil in their experience (and the reason for occasional moves to unbelief). However, the standard response does not seem adequate for Ellis or his contemporary fellow travelers such as Russell, Haught, and Peacocke. As we have seen, the response of modern theology to the problem of evil is to answer Ellis's question by saying that God *does not intervene in nature at all*. This is what "hands-off theology" means: God never intervenes directly, never breaks or overrides the laws of science, but lets them run their course. Natural evil can therefore occur according to the run of scientific law, and because of the mixture of law and chance, including in the process of evolution.

We need to remind ourselves that these thinkers do not hold that the reason God does not intervene to prevent natural evil is because he does not exist. Nor is it that, even though he exists, he is *unable to intervene*, since, as theists, they hold that God created the world and has the power to intervene in it directly (even though, according to most of them, he does not). Nor do they think that miracles (interventions by God directly in nature) *cannot* occur because they are impossible. Nor that the biblical miracles could not have occurred and that the Bible is best read as a book of metaphors that illustrate deeper spiritual and moral truths. These would all be atheistic arguments based on an appeal to the denial of God's existence, among other things. No, their main reason to support the view that God does not intervene in his creation is an appeal to modern science. They are impressed by the argument that we can always find natural explanations for events in the world, even apparently miraculous ones. This is a type of inductive argument based on the advancement and success of science. So they rule out, almost as a matter of policy, the possibility that God ever intervenes directly in nature. They then seek alternative ways to explain how God could affect nature without intervening, or without breaking any laws, as we have seen.

So their overall perspective is based mainly on an argument that appeals to the success of scientific explanation. However, there is another separate underlying logical argument that they do not always bring out, but that is revealed clearly in Ellis's question, and in his follow-up concerning why God does not intervene in those cases where it seems most obvious he should be intervening, that is, to prevent suffering that seems to us unfathomable (and frankly inconsistent with God's nature). Ellis's answer is that it is best to conclude that *God does not intervene at all in any direct way*; he thinks that if we adopt this view, our worry about evil *dissolves*. We also noted earlier that some go further and suggest that this way of thinking about the matter is actually a *gift* to human beings because it helps us deal better with the problem of evil.

Are these persuasive arguments? Do they advance a viable position on this vital question? We must acknowledge that they are serious responses. Even if it is true (as has been suggested) that many of these theologians are intimidated by science, we must concede that the scientific argument is not one that we can dismiss easily. The logical argument, too, is quite commanding, and we have already recognized that it has given many religious believers pause, though if it moves people to change radically their views, it seems more likely to push them toward unbelief rather than to the theological position that God does not, in fact, intervene directly in nature.

How seriously should we take the scientific argument? Theologians such as Bultmann have laid it down almost by fiat that we cannot believe in miracles (i.e., God's intervention) in the light of modern science. Haught's position is very close to this. However, there are problems with this point of view. For one thing, it overlooks the fact that for theists, science is not the whole story about reality. Theists believe that God created the world, its ingredients and laws, and structured them in a certain way, let us recall. So Bultmann's approach seems to rule out arbitrarily the fact that God could intervene in scientific law if he wished. Bultmann might insist that he is ruling it out based on scientific evidence. Yet, second, this move seems illegitimate since scientific evidence, while it cannot be ignored when considering this question, does not settle it definitively one way or the other. Scientific evidence is consistent with allowing for occasional miracles. The question is whether the logical step from the fact that there are not that many (apparent) miracles, or that some events have been later explained by natural causes, to the conclusion that they never occur is justified. Most theists believe it is not, especially since this conclusion would also have the implication that we would then need to reinterpret central religious practices, such as prayer, as well as the traditional view of God and much of biblical revelation. This is a high price to pay for what seems like a rushed and speculative conclusion, a theological concession many regard as a bridge too far. Of course, the atheist would insist on the stronger conclusion, that all of this shows that God does not exist, but we are not debating with the atheist here. Both sides in the current dispute believe in God as creator and sustainer and as immanent in the world. This is why the further claim from some of these theologians that God still intervenes in some way, is immanent in his creation, even though he never suspends scientific laws, does seem like having one's cake and eating it, or, to put it more bluntly, is a view that it is very difficult, perhaps impossible, to develop in a consistent way.

What about the logical side of the argument advanced by Ellis and others? One can see why one might be tempted toward the view that

God never intervenes, even if one were not totally subservient to modern science. Nevertheless, the price is still high and would require a radical rethinking of our view of God in a way that many will judge not viable. It moves theism much closer to atheism, it seems, without actually being atheistic, especially if one embraces the more radical version that the universe is governed by chance, that God never suspends or overrides the laws of physics, and that God remains ignorant of the final outcomes of his creation. Most theists regard this as a non-viable position, one that would lead to a dead religion that rather leans into an atheistic perspective. But the more serious problem here is that this response does not help us with the problem of evil. Haught, Ayala, Peacocke, and others have gone as far as to say that evolution in particular is a gift to theology, a gift from God to us. How should we understand the nature of the gift? The basic argument is that, in the words of Ayala, "a major burden was removed from the shoulders of believers when convincing evidence was advanced that the design of organisms need not be attributed to the immediate agency of the Creator, but rather is an outcome of natural processes. If we claim that organisms and their parts have been specifically designed by God, we have to account for the incompetent design of the human jaw [and other less-than-perfect designs]."[9] Haught welcomes the possibility of chance, evident he believes in evolution, because it allows the universe to "possess a strain of indeterminism, a potential for emergent freedom," thus allowing us to avoid the "straightjacket" of determinism and embrace an independent universe given to us by God in his generosity.[10]

Chance is an important part of Ayala's thesis. This is because he wishes to claim that the course of nature, with its imperfect designs, imperfections that lead to hardship, as well as the raging predatory battle among species and organisms, and so forth, are all a consequence of a significant element of chance operating in natural processes. Therefore, we cannot say that nature was designed (though, as we have seen, most of these thinkers still believe that God designed nature *overall*). This is clear in Ayala, who puts it this way:

> It is possible to believe that God created the world while also accepting that the planets, mountains, plants, and animals came about, after the initial creation, through natural processes. In theological parlance, God may act through secondary causes.

9. Ayala, *Darwin's Gift*, 159–60.

10. Haught, *Responses to 101 Questions*, 114.

> Similarly, at the personal level of the individual, I can believe that I am God's creature without denying that I developed from a single cell in my mother's womb by natural processes. For the believer, the providence of God impacts personal life and world events through natural causes. . . . [S]cientific conclusions and religious beliefs . . . do not stand in contradiction.[11]

Earlier, he suggested that "Science, much to the relief of theologians, provides an explanation that convincingly attributes defects, deformities, and dysfunctions to natural causes."[12] We noted earlier that it is not really evolution that is the gift for these thinkers, even though this is how they express the point. It is the fact that the universe operates indeterministically with a large element of chance that is the gift, for it is *these features* that make it possible to claim that the origination of the species, along with all of their imperfections, is due to chance and so is not designed. Indeterminism and chance make their particular take on evolution possible. God's non-intervention is also essential here. Although conceptually distinct from the notion of chance (since God could still intervene in a world full of chance) they go together in this argument because if you claim that evil occurs by chance, and that God never intervenes in his creation, then you can better explain why evil exists. Inaction by God in our world seems to lead to God's innocence with regard to the occurrence of evil, they suggest.

However, there is a clear problem facing this general response to the existence of evil. If God deliberately created nature with a significant degree of chance, and never intervenes directly in nature, isn't he still responsible for the evil that occurs as a result of this chance? This is an especially difficult problem if one also holds that God knows the future. *Ayala does not raise either of these questions.* Nor does Haught. These thinkers believe that if you say that God decides never to intervene directly in creation (as we saw with Ellis), then you can say that this is why God does not stop evil. Moreover, since there is a large element of chance in nature, this is also why a lot of bad things happen that were not designed, deliberately planned, or built into the initial blueprint. However, the problem surely is that even if we say that God has pledged never to intervene, and this might explain why God does not prevent evil from occurring, it would *not* explain why evil occurs in the first place, why such evil is present in creation. And so it would not follow that God

11. Ayala, *Darwin's Gift*, 175.

12. Ayala, *Darwin's Gift*, 156.

is not responsible for the evil that occurs, because he could have set up a different world in which he never intervenes, but that also has no, or very little, evil. Or he could intervene more regularly, even in a world of chance, and so this question must still be faced. Collins and Giberson reply to this criticism by arguing that God has put a certain freedom into the world, both in how it behaves naturally and also through human decision-making. However, they fail to see that, while the appeal to human freedom as support for the freewill defense may work for moral evil, it does not work for natural evil. As we noted, by "freedom" they appear to mean chance, but this does not explain why God put chance that leads to evil into the world in the first place, let alone how we would get purpose out of a world that is influenced by chance.

While acknowledging that this answer still does not explain why God would allow the holocaust, and that it just pushes us back to the problem of evil in general, Collins and Giberson argue that if God interfered regularly in nature to prevent natural evil, the world would be an inconsistent chaos, and science would be impossible.[13] This is an interesting point, but I don't think it is quite accurate, because, as we mentioned in the previous chapter, God could still intervene occasionally without causing too much disruption, so that the regular patterns (laws) that science requires are not compromised. Let us recall Murphy's point that God does not intervene *too often* because we need a reliably operating universe, but it does not follow that he *never* intervenes. Murphy suggests this as a reason for why God *never* breaks natural laws to prevent evil, and it also has echoes of Peacocke's suggestion that God suffers with us. However, this answer is also problematic for it seems to be saying the following: (A) God created X with properties that can cause suffering; (B) God decides to "cooperate" with X but not override the problematic properties so as to maintain consistency; so: (C) God is not responsible for the suffering caused by X.[14] This approach fails for the simple reason that God could have chosen not to have created X in that way in the first place. The crucial question surely is this: Is God really any less responsible for evil if we say that he created a world full of chance that leads to evil and in which he never intervenes, rather than a deterministically operating world that has evil in it, and in which he seems to intervene only "capriciously" (in the language of David Bartholomew)?[15] Creating

13. See Collins and Giberson, *Language of Science and Faith*, 140.

14. See Murphy, "Divine Action in the Natural Order," 355.

15. See Bartholomew, *God of Chance*, 25.

a universe with a lot of chance that leads to evil and in which God never intervenes seems to me no less of a problem than creating a universe where everything is designed to occur and only intervening occasionally. Either way, you have to face the problem of evil. Yet, our approach in this book is best in terms of the related and unavoidable question of purpose.

We should also consider whether evolution makes the problem of evil worse. This question is often raised because evolution reveals that evil is built into the very heart of creation, and is not just a byproduct of an otherwise good process. The evil and suffering that appear to be at the heart of nature raise the problem to a new level, it is argued. However, I am not convinced by this suggestion. Even before evolution, we knew about suffering and death in nature, about extinctions, about predation, and about the great hardship in nature and in life. The imperfections and flaws in the way organisms form seems to be of the same kind as noting that the operation of the heart is not perfect because the arteries can become clogged, or that hip joints are not perfect because they wear out, or that, more generally, biological organisms are not perfect because they are subject to disease. These are all problematic features of nature that force us to raise the problem of evil in the first place. No theist wishes to deny this, or to downplay the fact that natural evil gives rise to a legitimate question about God's existence and action. Yet, the evidence from evolution does not make the problem any worse than it was before. Evolution reveals a few more of the imperfections and flaws in nature, but this seems to be akin to the discovery of emerging diseases, rather than fundamentally changing the nature of the problem. Saying, as Ayala and Collins do, that if God had designed the eye, he wouldn't design it the way it is seems to me to be equivalent to saying that he wouldn't design the world the way it is. We can include the imperfections in nature as part of the general problem of natural evil; there seems to be no difference in kind between saying that God allows disease or a tsunami and that he permits less than optimal structural developments in biological organisms.

And we might add that the discovery of the process of evolution might make the problem of evil a bit easier to handle, but not in the way that Ayala and others suggest! It is part of the general way that human discovery and progress in knowledge help to reduce suffering. The more we discover about how things operate, the more we can reduce suffering. Indeed, today the line between natural and moral evil is becoming blurred in some areas as we gain more control over nature. We can see this if we compare, for instance, the number of deaths from famine in the tenth century with the number of deaths from famine in the twenty-first century. In earlier times, we could blame the evils caused by famine on events beyond human control, but today this response is no longer adequate. We would need to hold ourselves morally responsible, since we have the know-how, the wealth, and the technology to prevent starvation from famine. So evolution, especially when coupled with genetics, fits into this general pattern of human progress. The more we know about the development of life in its various aspects, the more we can reduce suffering. In this way, many new scientific discoveries help to reduce the amount of natural evil. This too is yet a further indication of the determinism that operates in nature, since finding the cause of a destructive genetic mutation, for example, and then taking steps to prevent it, requires that we reject the view that nature operates by chance.

Human Beings on a Different Plane

Humanity created in the image of God, and thus the special place of humans in God's creation, has been a foundational theme of Christian theism. However, it has been challenged by evolution, which suggests that human beings come out of the same line as all other species, a fact that many believe (both supporters and critics of evolutionary theory) diminishes our status in the tree of life in a significant way. While more advanced, *Homo sapiens* as a species is only further along on the continuum of living things rather than being in a separate category. This is especially true, atheistic naturalists will argue, when we take into account macroevolution, the claim that all species are genetically connected to each other (a claim supported by DNA evidence), along with the fact that *Homo sapiens* and other higher primates, such as apes and chimpanzees, share a common ancestor. This argument is also at least partially embraced, as we noted, by some of the alternative theorists of theistic evolution we considered above, especially those who espouse a strong view of the role of chance in the universe. Thinkers who hold this view believe that the existence of the human species was not planned by God but is an accidental outcome of a chance process; so it seems harder on such a view to position *Homo sapiens* on a different plane to other species.

These considerations raise the question of whether human beings differ in kind from other species or only in degree. A species might be said to differ in kind if it has fundamentally different properties than those with which it is compared. These properties must be on an entirely different plane and not just be different in degree or in quantity. They must be such as to place the entity in a different category by virtue of its very essence or nature. Centuries of Christian thinking have held that humanity does indeed differ in kind because of our unique properties, which include higher consciousness, reason, free will, moral agency, and the rational soul, among others. Many religious denominations affirm that humans are the only creatures that God created for his own sake, and most also teach that the human soul is specially created by God and is not the product of material processes. There are a variety of views on the nature and origin of the soul, and on its relation to human consciousness, which is usually understood as a part of, but not necessarily identical to, the soul.[16] Overall, the unique properties of humans are an unusual feature of the universe and are the basis of an argument for design. These unique qualities, I suggest, not only place human beings in a different category, but also add significantly to the viability of our view of theistic evolution. This is because the most reasonable explanation for their existence is that they were intended or planned by an intelligence. As I have tried to show throughout, the argument that they came about by chance is not believable, whether it is advanced by atheistic naturalists or Christian theologians.

We don't have space here to defend in detail the view that human beings differ in kind from other species, and in any case, these arguments have been discussed extensively in the past two decades, and I commend that fascinating topic to readers.[17] However, let us summarize a few main reasons for why any other conclusion is difficult to accept. First, the evidence that humanity is in a different category is overwhelming. This evidence consists not only of the general progress and achievements of humanity, but also of our quite remarkable human qualities (even leaving aside the question of the soul), such as consciousness and sense of selfhood, rationality and logic, free will, and other related features of human life, including the phenomenon of truth, and our pursuit of objective knowledge (including with regard to the moral order, of which more in

16. We don't have space for this fascinating topic here; for further discussion, see Aquinas, *Questions on the Soul*; Green and Palmer, *In Search of the Soul*; Murphy, *Bodies and Souls, or Spirited Bodies?*; Cooper, "Human Nature"; Swinburne, *Are We Bodies or Souls?*

17. From among many works, see MacIntyre, *Dependent Rational Animals*; Taylor, *Sources of the Self*; Kass, *Life, Liberty, and the Defense of Dignity*.

a moment), the consistency of mathematics with reality, not to mention such defining human experiences as spirituality, love, wonder, compassion, justice, and many others. These phenomena have proved almost impossible to account for under any naturalistic theory, which is why so much work today in philosophy and science simply *assumes* a naturalistic explanation as a starting point and goes from there to speculate about what might be the case or simply ignores the problems altogether.[18] The philosophy of mind and language is dominated by these assumptions, as is much work in psychology. And we must stress that the theory of evolution itself does not show the coming into existence of this kind of complexity in a purely natural way—that is, in a way that excludes design. As we saw earlier, this is mainly an assumption of the theory. It is more reasonable that our special human qualities are an intended result of the evolutionary process, rather than a random accident.

The existence of these traits leads to our second point. The phenomenon of higher beings with self-consciousness is a truly unique development, one we take too much for granted. It is really extraordinary that beings have evolved who both understand and have significant control over the process out of which they have come. This is surely *an enormously significant occurrence*, a further argument that we differ in kind from other species. We now seem to be at that stage where we have so much control over evolution that one might be tempted to conclude that the process will essentially stop with us; that we will remain at the top of the tree of life because we will be able to prevent our ever becoming extinct (and so no higher species will evolve from us). It is now highly unlikely that the human species will ever become extinct by the ordinary process of evolution on earth. This is a further consideration in favor of the view that human beings are intended by God to be at the top of the tree, and have been given intentionally the ability to understand and to control the process of evolution. From a moral point of view, it is also remarkable that we have developed and are continually developing a moral order (especially since the process is supposed to be non-moral and *value-free in itself*). We consult this moral order to make judgments about the outcomes of the process, such as that life is extremely valuable, that animals and indeed all species have a certain worth, even that the

18. A good example is Crick, who relegates the question of free will to a postscript; see his *Astonishing Hypothesis*, 265–68. Ignoring, rather than confronting, difficult problems is now a standard approach in the philosophy of mind, with many works excluding immaterial explanations by fiat.

impersonal environment itself has value. These kinds of moral judgments speak against chance being at the heart of nature, and move human beings into a different category.

A third reason is the fact that human beings have free will. We defined free will earlier as the ability to make a genuine free choice between alternatives, a choice that is not caused in the scientific sense. Here, we only need to emphasize the fact that it is quite astonishing that we have a quality that is not accessible to scientific investigation and, indeed, that seems to resist scientific study by its very nature. This property is also, of course, the basis of moral responsibility, punishment, and political freedom. Human life is unthinkable without free will, and those theorists who believe that it does not exist, such as Edward Wilson and Jerry Coyne, are simply barking up the wrong tree. From a theoretical point of view, the denial of free will has proved impossible to defend, and it is also a conclusion that is impossible to accept from a practical point of view. One reason these thinkers propose such an odd view is because they cannot see any place for free will in a universe that is purely material, one that is without purpose and was brought about by a haphazard process dominated by chance. But we can just as easily use the phenomenon of free will as a kind of *reductio* argument against an atheistic, materialist, purposeless view of human existence. Descartes's way of putting it is very apt: Free will exists, he observed from experience, and so it follows that materialism must be false.[19] Moreover, even if one denies that free will is real, one must still act *as if* we have free will (this is why its denial is ridiculously impractical). One must also hold that, from the point of view of truth, it is wrong to reject its denial (a stance which also requires the acceptance of free will!). So the consequences of its denial are unsettling (hence the *reductio* argument in its favor).

The phenomenon of free will also coheres extremely well with the Christian injunction to live a good moral life, as well as making sense of notions such as redemption, forgiveness, compassion, and the pursuit of justice, among many other human activities. All of these essential human experiences are made possible by the fact that we can respond freely to other persons and the nature of events, another indication of our difference in kind from other species. Contemporary theories of theistic evolution also run into problems with regard to many of these issues. Thinkers such as Haught, Coyne, Collins, Ayala, and others don't seem to

19. See Descartes, *Meditations* (esp. the Sixth Meditation); also *Principles of Philosophy*, esp. Part I, Sections 7, 39–41.

be able to affirm consistently that human beings differ in kind from other species, given their acceptance of an evolutionary process that operates by chance. It is hard to see how these thinkers can argue that the qualities we have discussed here—such as reason and logic, free will and moral agency, even higher-level consciousness—can be in any way necessary if they emerged from a chance process. It is a central belief of the Christian view that these qualities are intended by God, in which case they are necessary and had to come out of the evolutionary process. It is not clear that Haught or Ayala could hold to this view, given the dominant role they assign to chance throughout the course of nature. This is quite a serious problem for them since it seems it would have to extend to other features of humanity, such as objective morality and human culture, issues to which we will now turn.

The Foundation of the Objective Moral Order

One of the major problems facing (especially) atheistic views that appeal mainly to evolution is how to justify the objectivity of morality. Objective morality refers to the idea that moral values are objectively true in themselves and not dependent on human opinion, attitudes, feelings, or cultural beliefs. Belief in objective morality is a bedrock commitment of the human species. Even though we disagree about the content or range of values that make up the moral order and, over time, have changed our beliefs about what properly belongs to this order, we agree that *there is an objective moral order, which we are obliged to follow*. Moreover, we organize our societies around this moral order, critique other societies if they do not agree, and punish those who violate the objective order, sometimes very severely. The problem is that it is difficult to *justify* this moral order if the theory of evolution is one's main foundational argument for atheism. If there is no designer and no overall meaning or purpose to the universe or to life (no teleology), then it is challenging to establish an objective moral order on a firm foundation. The necessity that is required to justify it is precluded by the way it originated, just as it is for other phenomena such as human nature or essence, differences in kind between species, gender differences within species, and so forth. How can one establish an objective, necessary foundation for a moral order that emerged out of a purely physical, natural process that in itself is neither moral nor immoral, and has no purpose behind it, was not designed and is not aimed at any particular goal?

This question of justification gets to the heart of the matter. It is a deeper question than the question of whether we currently believe in an objective moral order (we clearly do) or whether we act on the basis of objective morality (again, we clearly do); the problem we are focused on is how this moral order is justified in any objective sense, given its origins. Some thinkers acknowledge this problem, but understandably shrink from the inevitable consequence of undermining the moral order. One such thinker is Edward Wilson, who has pioneered the field of sociobiology, the study of the biological basis of social behavior, using evolutionary principles as a guide.[20] Although Wilson's view has met with a lot of resistance, it is surely the logical next step for evolutionary atheism. While acknowledging the obvious point that genetics, environmental factors, and human culture play some role in human decision-making (but not free will, which he was inclined to deny, suggesting it was likely a byproduct of brain complexity), Wilson advanced the radical proposal that human beings are shaped mostly by their genetic inheritance and their environments.[21] So all products of human culture, including morality, are mostly influenced by genetics. Moreover, he went further and argued that moral behavior must be explained in terms of its evolutionary advantages. He has been led to this view because of his faith in the power of evolution to explain even those aspects of our nature that do not appear to be part of the physical realm. We need to acknowledge, Wilson believes, the role that natural selection must play in the origin of our social behaviors, such as group kinship, community, competition, altruism, and so forth. These behaviors, and indeed all of human culture, evolve just as our physical traits do, and like those traits, they are aimed at enhancing survival and reproductive advantage. Nevertheless, Wilson does not then conclude that we should abandon the idea of objective morality. Rather, he suggests we adopt a pragmatic approach that emphasizes values that aid our survival and physical well-being!

Needless to say, many thinkers, such as Gould, who are on the same side as Wilson in the broader debate about the serendipitous origins of these phenomena, have critiqued Wilson's general approach because they believe it could be used to justify discrimination, and that would be

20. See Wilson, *Sociobiology*; for a critique, see Rose et al., *Not in Our Genes*; also, Rose and Rose, *Alas, Poor Darwin*.

21. See Wilson, *Sociobiology*, ch. 1; Wilson, *Consilience*, 130–32.

immoral![22] But this consequence is one that Wilson is honestly acknowledging. If, as he thinks must be the case, given his starting perspective based on chance and lack of purpose, all of our traits evolved according to the process of natural selection (a view that Gould accepts), then it must be the case not only that our morality evolved, but that it could have evolved differently. This is similar to Gould's argument that if we ran the tape of life again, we might get no species of *Homo sapiens* at all. Gould is simply shrinking from the consequences of his own argument. Indeed, thinkers like Dennett and Dawkins have stridently tried to account for *religion* in the same manner, suggesting that such beliefs evolved and survived because of the adaptive advantage they accord to humans in the struggle for survival.[23] So why not offer the same explanation for moral beliefs, and draw the same conclusion—that they are false, and we should abandon them?! The problem is severe because it seems to mean that we cannot rationally regard our current moral order as necessary and objective. One could still argue that we should behave in accordance with it since it is the order that has evolved, but once one undermines the theoretical basis for morality, it becomes harder to adhere to moral standards (as the influence of moral relativism in the twentieth century amply illustrates). Moreover, the actual origins of our moral values are shrouded in mystery if one accepts an evolutionary explanation. Not to mention the fact that there are many values that contradict a reductionist evolutionary explanation, such as self-sacrifice, the Christian understanding of love, altruism, love of neighbor, among others. Motivating human behavior that is both admirable and inspirational, all such values belie an evolutionary origin. Theistic evolution is a far more convincing explanation, it seems, since it affirms the objectivity of the moral order, one that was designed by God.

This argument may also be stated in the form of a *reductio*. The reasoning is straightforward: If evolution occurred by chance, then moral values are arbitrary, yet far from being arbitrary, they are objective and necessary; therefore, evolution occurred by design. It is not so much that we cannot accept that morality is arbitrary, for moral behavior is a defining part of the very essence of human beings, and so we would have to commit to a deeper implication, that our essential nature is also arbitrary. It also follows that if morality is arbitrary, it would be challenging to

22. See Gould, *Ever Since Darwin*, ch. 32; also, Gould, *Mismeasure of Man*.

23. See Dawkins, *God Delusion*, ch. 5; Dennett, *Breaking the Spell*.

justify various forms of punishment for immoral actions (except perhaps psychological and social forms).

Atheistic evolutionary accounts also flirt seriously with moral relativism for several reasons. One is that it is obvious that the moral order that emerges from a process governed by chance must be contingent, not necessary. Second, different moral orders could evolve in different populations (and, we must add, have done so) or at different paces in history, and so it is deeply problematic to impose values that have emerged from one evolutionary pathway onto other pathways under the guise of "moral objectivity." Third, it seems to follow that, as with many naturally occurring processes (e.g., relating to illness, appearance, etc.), we do not have to stick with what nature has provided, but can change the moral order to suit our current (subjective) tastes and interests. All of this seems to follow from any view that claims that the evolution and content of the moral order is a chance occurrence.

The strength of the theistic view is obvious here. The Christian view holds that God is the author of the objective moral order that has emerged through the evolutionary process in human experience. It was intended by God, is founded upon God's nature and so is necessary, and is a central part of the story of God's purposes for humanity. This philosophical view also complements the biblical view that humans are made in God's image, and one recalls Thomas Aquinas' argument that the moral law is written on the human heart. It is extremely difficult to accept that the objective order that we live by and strive to perfect, and that contributes so much to the formation of our character and sense of self, as well as shaping our social and political lives, is no more than an accident of nature, as indeed is all of human culture. The Christian view makes more sense of all of these issues. Even if this argument does not lead one to embrace a religious view of reality, it does clearly show, I contend, that theistic evolution is at least viable!

Meaning and Purpose

A consuming concern of human life—perhaps *the* central question for most people—relates to the question of meaning and purpose. And it is on this question that a viable theory of theistic evolution is far superior to atheistic and materialist alternatives. Like some of the issues above, this matter can be considered on its own, or as part of an overall case, where each argument is reasonable in itself, but when taken altogether, the arguments reinforce each other with a cumulative effect to present a very strong—certainly a viable—case for Christian theism. And it is a Christian theism that takes full account of modern science, including evolution, as we have seen.

The question of meaning should not be underestimated in the modern world. There are at least two important aspects to this question today. The first is the traditional religious approach, which is that human beings understand their lives and experiences to have a deeper purpose, some overall meaning and direction, even if all facets of it cannot be completely discerned to our complete satisfaction. But this purpose gives meaning to our deepest experiences such as love, moral behavior and ideals, family relationships, community and social relations, religious transcendence, ideals of culture and civilization, the pursuit of knowledge and truth, and our deepest expressions of ourselves in art, literature, music, and other cultural activities. Our individual, localized forms of living, in family and community, find their foundation in an underlying layer of meaning (even if we occasionally lose sight of it or do not give it enough attention). We believe that we have been given the gift of life by God for a reason and that we are obliged to live the best moral life we can according to an objective moral order, given and even despite our circumstances, and that we will enjoy eternal life with God, who created the universe for this purpose. We recognize that suffering is part of our journey, but hold a reasonable hope that the promise of eternal life through Jesus Christ is the ultimate destiny of humanity.[24]

The second aspect is that today, perhaps more than ever before in history, there is a quite noteworthy *loss of meaning* that afflicts modern living. This is not unrelated to the waning or downplaying of religious views of reality, with the attendant loss of an overall significance to human activities. Irrespective of one's religious denomination or religious morality, one thing common to most of the world's great religions is that they provide meaning and an overall rationale to our lives. This is one of the features that gives the religious way of life its appeal, and why so many find it profoundly satisfying (indeed, those who are not religious often envy this aspect of religion as they lament the lack of meaning in their situation).

24. For detailed, compelling accounts of the Christian philosophy of life, see (among many others): Williams, *Tokens of Trust;* Stump, *Wandering in Darkness*; Willard, *Divine Conspiracy*; Wright, *Simply Christian*; Ratzinger, *Christianity and the Crisis of Culture*; Masterson, *In Reasonable Hope.*

Our contemporary times have been significantly defined by the skeptical and cynical tendencies that have ushered in a loss of meaning, despair, and even a certain anarchism and nihilism in many cultures (a frequent subject and indeed influence in contemporary literature, art, and film). A rationally defensible religious view, such as the one defended in this book, often serves as an antidote to this modern malaise. There are many destructive tendencies that emerge from secularist alternatives, especially loss of meaning and purpose, and many societies are under severe pressure from the weight of them. This can be seen in modern trends that are expressed in dysfunctional relationships, drugs and escapism, consumerism as an avocation, increased levels of street and other crimes, mindless violence, a mental health epidemic, disturbing levels of addiction, social unrest, and inclinations toward class conflict, despair, and loneliness, among other pathological tendencies. Secularist and atheistic alternatives to religion are, after all, untested, and the initial results are not promising, although not everything in these views is destructive, of course. Nor do I wish to say that a secularist worldview in itself is responsible for these problems; the main point is that the loss of meaning, which is characteristic of that approach, for many naturally finds expression in many of our modern personal and social problems. Organized religions, political institutions, and other institutions of the state also must accept their share of responsibility in contributing to the destructive mindset that has taken root among many, especially in Western countries. Our main point, however, is that looking at the religious worldview from the point of view of its positive features, as a viable philosophical and theological approach, as a philosophy of life, one can clearly see that in ordinary practical life it offers a more satisfying and productive theory of meaning and purpose than secularist alternatives based on evolutionary naturalism.

It is important to consider the question of meaning not only as an experiential question, where we come into contact with it the most, but also from a more philosophical point of view. This can be brought out clearly when we consider the nature of the universe we live in. The sheer size of the cosmos is staggering. Our own galaxy, the Milky Way, which consists of most of those stars we see in the night sky, is around one hundred thousand light-years across, and at least one thousand light-years thick (recall that a light-year is the distance light travels in a year, nearly six trillion miles). The Milky Way contains at least two hundred billion stars. It is currently estimated that there are billions of such galaxies that make up the universe, and the earth, on current knowledge, is the only one we know of that has life. When contemplating the nature of this universe, there are at least three points that stand out to an inquiring mind: (1) What we may describe as the *majesty* of the universe: its sheer size, its structure, its lawlike operation and order. (2) Its *beauty*: the aesthetic dimension, along with the wonders of existence and of life, including conscious life, facts that hold enormous appeal and cannot be disregarded. (3) Its *question*. The latter is the most central matter from the philosophical point of view and relates directly to the question of meaning. We want to know how it got here, what is the purpose behind it, and what is our purpose in it (among other related questions). Although some have tried to ignore the question the universe poses, or to dismiss it, or even act like it has no significance for them, it is hard not to regard such reactions as attempts to avoid the issue. The universe's "question" is reasonable, essential and, above all, must have an answer. Perhaps, we might even say that whether or not it has an answer in itself, it is one that human beings *must* answer! From our growing knowledge of the universe, we know that it did not cause itself, and must have some ultimate cause. As noted above, whether one accepts theism or not, that much is true; it is also why I believe that no atheistic view can work—because it cannot in principle answer this question.

So the question of meaning is a key part of the human condition, and it fills the universe with significance. Otherwise, the universe would be just there, large, unreceptive, and without purpose or direction. And the significance of the question is real, not manufactured. Does it follow that if we crave an answer, there must be one? No, but it is the most reasonable perspective to adopt, given all we know, including from evolution and from human progress, a phenomenon not to be underestimated. It is clear from evolution that we started in very humble, primitive conditions. And yet after the emergence of *Homo sapiens* and the eventual development of higher consciousness, language, logic, and intelligence, along with our moral nature, our progress is nothing short of astonishing, especially when considered in its totality. From the basic inventions, tools, and social arrangements of primitive man, through the stone age (approximately three million years ago), to the bronze age (five thousand years ago) and iron age (three thousand years ago), through classical antiquity, into the Middle Ages, through the Renaissance and the Enlightenment periods, up to present times, it seems clear that the human race is on a journey. Although not the shortest distance between two points by any means, we have advanced from primeval, dangerous, squalid conditions when meaning and purpose surely often seemed elusive to more sophisticated, safer, healthier societies and cultures, and have made remarkable progress in politics, morality, and the institutions of civilization in general (of course, we still have a long way to go, but this point does not distract from the fact that human progress is not just impressive, but mind-boggling). Developments in such areas as technology, communications, travel, medicine, education, human rights, the developing political state, and with regard to the environment, have also transformed modern life (while acknowledging, of course, that not every new idea, trend, or innovation is for the better).

Given our humble origins, our success in making sense of and in managing our lives and the world demands attention as a phenomenon in itself, let alone what the future may hold as the human race continues to advance. Given our extremely primitive origins, our journey to where we are now suggests a deeper meaning to human existence. Human beings can write symphonies that express the essence of the human spirit, can build computers, engines, and airplanes, understand the past (including our own origin), develop world communication systems, perform impressive feats in medicine, explore space, practice philosophy, science, mathematics, music, and art, reach toward God and infinity, and achieve inspiring acts of reasoning, understanding, love, sacrifice, and moral integrity. Given our origins in the wild distant past, and our early savage nature, these developments are not just surprising but almost incredible.

The nature and achievements of humanity suggest two things, according to the theist. First, that there is an overall design to the universe in which human beings are most likely the intended outcome. And second, that human beings have a higher purpose or *telos* within God's creation. Even with our evolutionary origins, this is the most reasonable position to take with regard to the question of meaning and purpose. It is a clear reason, I suggest, for why theistic evolution (especially the version defended in this book) is not merely viable but is the most viable position with regard to the ultimate questions of human existence.

Bibliography

Alexander, Denis. *Creation or Evolution: Do We Have to Choose?* Oxford: Monarch, 2008.

———. *Is There Purpose in Biology?* Oxford: Lion, 2018.

Alston, William. "Divine Action: Shadow or Substance?" In *The God Who Acts: Philosophical and Theological Explorations*, edited by Thomas Tracy, 41–62. Philadelphia: Pennsylvania State University Press, 1994.

Aquinas, St. Thomas. *Disputed Questions on Truth*. Chicago: Regnery, 1954.

———. *Questions on the Soul*. Translated by James Robb. Milwaukee: Marquette University Press, 1984.

———. *Summa Theologiae*. Translated by Fathers of English Dominican Province. New York: More, 1981.

Aristotle. *Physics*. Edited by David Bostock. Translated by Robin Waterfield. New York: Oxford University Press, 2008.

Augustine. *On Genesis*. Edited by Boniface Ramsey. New York: New City, 2004.

Ayala, Francisco. "Darwin's Devolution: Design Without Designer." In *Evolutionary and Molecular Biology*, edited by Robert Russell et al., 101–16. Rome: Vatican Observatory, 1998.

———. *Darwin's Gift to Science and Religion*. Washington, DC: Henry, 2007.

———. "Darwin's Greatest Discovery: Design Without Designer." *Proceedings of the National Academy of Sciences* 104 (2007) 8567–73.

Ayala, Francisco, and Theodosius Dobzhansky, eds. *Studies in the Philosophy of Biology: Reduction and Related Problems*. London: Macmillan, 1974.

Baron, Craig. "God Is Deeper than Darwin." *Heythrop Journal* 54 (2013) 645–57.

Bartholomew, David. *God of Chance*. London: SCM, 1984.

Beatty, John. "The Evolutionary Contingency Thesis." In *Concepts, Theories, and Rationality in the Biological Sciences*, edited by Geron Wolters and James G. Lennox, 45–81. Pittsburgh: University of Pittsburgh Press, 1995.

———. "What Are Narratives Good For?" *Studies in History and Philosophy of Science* 58 (2016) 33–40.

Behe, Michael. *Darwin's Black Box*. New York: Free, 1998.

———. *The Edge of Evolution*. New York: Free, 2007.

Bohm, David. *Causality and Chance in Modern Physics*. Rev. ed. London: Routledge, 1984.

Born, Max. *The Born–Einstein Letters*. Translated by Irene Born. New York: Walker, 1971.

Bowler, Peter. *Evolution: The History of an Idea.* Los Angeles: University of California Press, 1989.

Brown, Janet. "Noah's Flood, the Ark, and the Shaping of Early Modern Natural History." In *When Science and Christianity Meet*, edited by David Lindberg and Ronald Numbers, 111–38. Chicago: University of Chicago Press, 2003.

Bultmann, Rudolph. *New Testament Mythology and Other Basic Writings.* Edited and translated by Schubert Ogden. Philadelphia: Fortress, 1984.

Carlson, Richard F., ed. *Science and Christianity: Four Views.* Downers Grove, IL: InterVarsity, 2000.

Cartwright, Nancy. *How the Laws of Physics Lie.* Oxford: Clarendon, 1989.

Charlesworth, Brian, and Deborah Charlesworth. *Evolution: A Very Short Introduction.* New York: Oxford University Press, 2003.

Cicero. *De Divinatione.* Translated by William A. Falconer. Cambridge: Harvard University Press, 1964.

Clayton, Philip. *God and Contemporary Science.* Grand Rapids: Eerdmans, 1997.

Collins, Francis. *The Language of God.* New York: Free, 2006.

Collins, Francis, and Karl Giberson. *The Language of Science and Faith: Straight Answers to Genuine Questions.* Downers Grove, IL: InterVarsity, 2011.

Cooper, John W. "Human Nature in Theistic and Evolutionary Perspectives." *Zygon* 48 (2013) 478–95.

Coyne, George. "God's Chance Creation." *The Tablet*, August 6, 2005.

Coyne, Jerry. "You Don't Have Free Will." *Chronicle of Higher Education*, March 18, 2012.

Craig, William Lane. *In Quest of the Historical Adam.* Grand Rapids: Eerdmans, 2021.

———. *The Kalam Cosmological Argument.* 1979. Reprint, Eugene, OR: Wipf and Stock, 2000.

———. "Response to 'Mere Theistic Evolution.'" *Philosophia Christi* 22 (2020) 55–61.

Craig, William L., and Quintin Smith. *Theism, Atheism, and Big Bang Cosmology.* Oxford: Clarendon, 1995.

Cushing, James. "Determinism Versus Indeterminism in Quantum Mechanics: A 'Free' Choice." In *Quantum Mechanics*, edited by Robert Russell et al., 99–110. Rome: Vatican Observatory, 2001.

Darwin, Charles. *The Origin of Species.* 1859. Reprint, London: Penguin, 1985.

Davies, Paul. *The Goldilocks Enigma: Why Is the Universe Just Right for Life?* Boston: Mariner, 2006.

———. "Teleology Without Teleology: Purpose Through Emergent Complexity." In *Evolutionary and Molecular Biology*, edited by Robert Russell et al., 151–62. Rome: Vatican Observatory, 1998.

Dawkins, Richard. *The Blind Watchmaker.* New York: Norton, 1996.

———. *Climbing Mount Improbable.* New York: Norton, 1997.

———. *The God Delusion.* New York: Houghton Mifflin, 2006.

———. *River Out of Eden.* New York: Basic, 1996.

Dembski, William. *No Free Lunch: Why Specified Complexity Cannot Be Purchased Without Intelligence.* New York: Rowman and Littlefield, 2002.

Dennett, Daniel C. *Breaking the Spell: Religion as a Natural Phenomenon.* New York: Viking, 2006.

Descartes, René. *Philosophical Writings.* Edited by John Cottingham. 3 vols. Cambridge: Cambridge University Press, 1985.

De Vries, Paul. "Naturalism in the Natural Sciences." *Christian Scholar's Review* 15 (1986) 388–96.

Dilley, Stephen. "How to Lose a Battleship: Why Methodological Naturalism Sinks Theistic Evolution." In *Theistic Evolution: A Scientific, Philosophical, and Theological Critique*, edited by J. P. Moreland et al., 593–631. Wheaton, IL: Crossway, 2017.

———. "Philosophical Naturalism and Methodological Naturalism: Strange Bedfellows?" *Philosophia Christi* 12 (2010) 118–41.

Dodds, Michael. *Unlocking Divine Action: Contemporary Science and Thomas Aquinas*. Washington, DC: Catholic University of America, 2012.

Dudley, John. *Aristotle's Concept of Chance*. Albany, NY: State University of New York Press, 2012.

Earman, John. *A Primer on Determinism*. Dordrecht: Reidel, 1986.

Ellis, George. "Ordinary and Extraordinary Divine Action." In *Chaos and Complexity*, edited by Robert Russell et al., 359–95. Rome: Vatican Observatory, 2000.

Erickson, Millard. *Christian Theology*. Grand Rapids: Baker, 1993.

Farrar, Austin. *Faith and Speculation*. Edinburgh: T&T Clark, 1967.

Ford, Joseph. "What Is Chaos, That We Should Be Mindful of It?" In *The New Physics*, edited by Paul Davies, 348–71. Cambridge: Cambridge University Press, 1989.

Freddoso, Alfred. "God's General Concurrence with Secondary Causes: Why Conservation Is Not Enough." *Philosophical Perspectives* 5 (1991) 553–85.

Geivett, R. Douglas, and Gary Habermas, eds. *In Defense of Miracles: A Comprehensive Case for God's Action in History*. Downers Grove, IL: InterVarsity, 1997.

George, Marie. "What Would Aquinas Say About Intelligent Design?" *New Blackfriars* 94 (2023) 676–700.

Goetz, Stewart, and Charles Taliaferro. *Naturalism*. Grand Rapids: Eerdmans, 2008.

Gould, Stephen J. *Ever Since Darwin*. New York: Norton, 1992.

———. *The Mismeasure of Man*. 1981. Reprint, New York: Norton, 1996.

———. "Sociobiology and the Theory of Natural Selection." In *Sociobiology: Beyond Nature/Nurture?*, edited by George Barlow and James Silverberg, 257–69. Boulder, CO: Westview, 1980.

———. *Wonderful Life: The Burgess Shale and the Nature of History*. New York: Norton, 1990.

Grant, W. Matthews. *Free Will and God's Universal Causality*. London: Bloomsbury, 2019.

Gray, Asa. *Darwiniana: Essays and Reviews Pertaining to Darwinism*. New York: Appleton and Co., 1876. https://www.gutenberg.org/ebooks/5273.

Green, Joel, and Stuart Palmer, eds. *In Search of the Soul*. Downers Grove, IL: InterVarsity, 2005.

Greene, Mott T. "Genesis and Geology Revisited: The Order of Nature and the Nature of Order in Nineteenth-Century Britain." In *When Science and Christianity Meet*, edited by David Lindberg and Ronald Numbers, 139–59. Chicago: University of Chicago Press, 2003.

Grudem, Wayne. "The Incompatibility of Theistic Evolution with the Biblical Account of Creation and with Important Christian Doctrines." In *Theistic Evolution: A Scientific, Philosophical, and Theological Critique*, edited by J. P. Moreland et al., 61–77. Wheaton, IL: Crossway, 2017.

———. "Theistic Evolution Undermines Twelve Creation Events and Several Crucial Christian Doctrines." In *Theistic Evolution: A Scientific, Philosophical, and Theological Critique*, edited by J. P. Moreland et al., 785–837. Wheaton, IL: Crossway, 2017.

Ham, Ken, et al. *Four Views on Creation, Evolution, and Intelligent Design*. Grand Rapids: Zondervan, 2017.

Harris, Sam. *The End of Faith*. New York: Norton, 2005.

Hasker, William. *God, Time, and Knowledge*. Ithaca, NY: Cornell University Press, 1998.

Haught, John, F. *Christianity and Science*. Maryknoll, NY: Orbis, 2007.

———. *The Cosmic Vision of Teilhard de Chardin*. Maryknoll, NY: Orbis, 2021.

———. *Deeper Than Darwin*. Boulder, CO: Westview, 2003.

———. *God After Darwin*. Boulder, CO: Westview, 2000.

———. *God and the New Atheism*. Louisville: Westminster John Knox, 2008.

———. *Mystery and Promise: A Theology of Revelation*. Collegeville, MN: Liturgical, 1993.

———. *The New Cosmic Story: Inside Our Awakening Universe*. New Haven: Yale University Press, 2017.

———. *Responses to 101 Questions on God and Evolution*. Mahwah, NJ: Paulist, 2001.

———. *Science and Faith*. Mahwah, NJ: Paulist, 2012.

Hebblethwaite, Brian, and Edward Henderson, eds. *Divine Action*. Edinburgh: T&T Clark, 1990.

Heisenberg, Walter. *Physics and Philosophy*. New York: Prometheus, 1999.

Hick, John, ed. *The Metaphor of God Incarnate*. Louisville: Westminster John Knox, 2006.

Hitchens, Christopher, ed. *The Portable Atheist*. New York: Grand Central, 2007.

Hodge, Jonathan. "Chance and Chances in Darwin's Early Theorizing and in Darwinian Theory Today." In *Chance in Evolution*, edited by Grant Ramsey and Charles Pence, 41–75. Chicago: University of Chicago Press, 2016.

Hoefer, Carl. "Causal Determinism." In *Stanford Encyclopedia of Philosophy*, edited by Edward N. Zalta and Uri Nodelman. https://plato.stanford.edu/entries/determinism-causal/.

Hume, David. *Enquiry Concerning Human Understanding*. Edited by Stephen Buckle. Cambridge: Cambridge University Press, 2007.

Humphreys, Paul. *The Chances of Explanation*. Princeton: Princeton University Press, 2014.

International Theological Commission. *Communion and Stewardship: Human Persons Created in the Image of God*. https://www.vatican.va/roman_curia/congregations/cfaith/cti_documents/rc_con_cfaith_doc_20040723_communion-stewardship_en.html.

Jaki, Stanley. *Genesis 1 Through the Ages*. Edinburgh: Scottish Academic, 1998.

John Paul II. "Truth Cannot Contradict Truth." Address to the Pontifical Academy of Science, October 22, 1996.

Johnson, Luke Timothy. *The Real Jesus: The Misguided Quest for the Historical Jesus and the Truth of the Traditional Gospels*. San Francisco: HarperOne, 1997.

Johnson, Phillip. *Darwin on Trial*. Downers Grove, IL: InterVarsity, 1993.

Kane, Robert. *The Significance of Free Will*. New York: Oxford University Press, 1998.

Kass, Leon. *Life, Liberty, and the Defense of Dignity*. New York: Encounter, 2002.

Koperski, Jeffrey. *Divine Action, Determinism, and the Laws of Nature*. London: Routledge, 2020.

Lamoureux, Denis. *Evolutionary Creation: A Christian Approach to Creation*. Eugene, OR: Wipf & Stock, 2008.

———. "Evolutionary Creation: A Christian Approach to Evolution." *The BioLogos Foundation*. https://cms.biologos.org/wp-content/uploads/2009/08/lamoureux_scholarly_essay.pdf.

Laplace, Pierre. *A Philosophical Essay on Probabilities*. 1814. Reprint, Hong Kong: Forgotten, 2012.

Lewis, Ard. "The Flawed Theology and Philosophy Behind Christian Resistance to Evolution." *BioLogos*, January 11, 2011. https://biologos.org/articles/common-concerns-about-the-implication-of-biologos-science.

Maar, Alexander. "What's a Chance Event? Contrasting Different Senses of 'Chance' with Aristotle's Idea of Meaningful Unusual Accidents." *Revista Archai* 32 (2022) 1–33.

MacIntyre, Alasdair. *Dependent Rational Animals*. La Salle, IL: Open Court, 2001.

Maitzen, Stephen. *Determinism, Death, and Meaning*. London: Routledge, 2022.

Masterson, Patrick. *In Reasonable Hope: Philosophical Reflections on Ultimate Meaning*. Washington, DC: Catholic University of America Press, 2021.

McCall, Bradford. *Macroevolution, Contingency, and Divine Activity: Divine Involvement Through Uncontrolling, Amorepotent Love in an Evolutionary World*. Eugene, OR: Pickwick, 2023.

McCarthy, John. "Letter to the Grand Duchess Christina: Justice, Reinterpretation, and Piety." In *Where Have All the Heavens Gone?*, edited by John McCarthy and Edmondo Lupieri, 22–41. Eugene, OR: Cascade, 2017.

McDonald, Patrick, and Nivaldo J. Tro. "In Defense of Methodological Naturalism." *Christian Scholar's Review* 38 (2009) 201–29.

McMullin, Ernan. "Cosmic Purpose and the Contingency of Human Evolution." *Zygon* 48 (2013) 338–63.

Meyer, Stephen. "The Difference It Doesn't Make: Why the 'Front-End Loaded' Concept of Design Fails to Explain the Origin of Biological Information." In *Theistic Evolution: A Scientific, Philosophical, and Theological Critique*, edited by J. P. Moreland et al., 215–36. Wheaton, IL: Crossway, 2017.

———. "Scientific and Philosophical Introduction: Defining Theistic Evolution." In *Theistic Evolution: A Scientific, Philosophical, and Theological Critique*, edited by J. P. Moreland et al., 33–60. Wheaton, IL: Crossway, 2017.

———. *Signature in the Cell*. San Francisco: HarperOne, 2009.

Mill, John Stuart. *A System of Logic*. Toronto: University of Toronto Press, 1974.

Miller, Kenneth. *Finding Darwin's God*. San Francisco: Harper, 2007.

———. *Only a Theory*. New York: Penguin, 2008.

Millstein, Roberta. "How Not to Argue for the Indeterminism of Evolution." In *Determinism and Physics and Biology*, edited by Andres Huttemann, 91–107. Panderborn: Mentis, 2003.

Mohseni, Masoud, et al., eds. *Quantum Effects in Biology*. Cambridge: Cambridge University Press, 2014.

Monod, Jacques. *Chance and Necessity*. New York: Knopf, 1971.

Moreland, J. P., et al., eds. *Theistic Evolution: A Scientific, Philosophical, and Theological Critique*. Wheaton, IL: Crossway, 2017.

Morris, Simon Conway. *Life's Solution: Inevitable Humans in a Lonely Universe*. Cambridge: Cambridge University Press, 2003.

Murphy, Nancey. *Beyond Liberalism and Fundamentalism*. Harrisburg, PA: Trinity, 1996.

———. *Bodies and Souls, or Spirited Bodies*? New York: Cambridge University Press, 2006.

———. "Divine Action in the Natural Order: Buridan's Ass and Schrodinger's Cat." In *Chaos and Complexity*, edited by Robert Russell et al., 325–57. Rome: Vatican Observatory, 2000.

———. "Phillip Johnson on Trial: A Critique of His Critique of Darwin." *Perspectives on Science and Christian Faith* 45 (1993) 26–36.

Murray, Michael, and John Ross Churchill. "Mere Theistic Evolution." *Philosophia Christi* 22 (2020) 7–41.

Nagel, Thomas. *The Last Word*. New York: Oxford University Press, 1997.

———. *Mind and Cosmos*. New York: Oxford University Press, 2012.

Paley, William. *Natural Theology*. 1802. Reprint, Whitefield, MT: Kessinger, 2003.

Peacocke, Arthur. *Creation and the World of Science*. Oxford: Oxford University Press, 2004.

———. *God and the New Biology*. San Francisco: Harper, 1986.

———. "God's Interaction with the World." In *Chaos and Complexity*, edited by Robert Russell et al., 263–87. Rome: Vatican Observatory, 2000.

———. *Intimations of Reality: Critical Realism in Science and Religion*. Notre Dame: University of Notre Dame Press, 1984.

———. *Theology for a Scientific Age*. Minneapolis: Fortress, 1993.

Peirce, Charles. "The Doctrine of Necessity Examined." In *The Essential Peirce*, edited by Nathan Houser and Christin Kloesel, 1:415–30. Bloomington: Indiana University Press, 1992.

Peters, Ted, and Martinez Hewlett. *Evolution from Creation to New Creation*. Nashville: Abington, 2003.

Phillips, D. Z. *The Concept of Prayer*. 1965. Reprint, Oxford: Blackwell, 1981.

Pinker, Steven. *Enlightenment Now: The Case for Reason, Science, Humanism, and Progress*. New York: Penguin, 2019.

Plantinga, Alvin. *Where the Conflict Really Lies: Science, Religion, and Naturalism*. New York: Oxford, 2011.

Polkinghorne, John. *The Faith of a Physicist*. Minneapolis: Fortress, 1996.

———. "The Metaphysics of Divine Action." In *Chaos and Complexity*, edited by Robert Russell et al., 147–56. Rome: Vatican Observatory, 2000.

———. *Quantum Theory*. Oxford: Oxford University Press, 2002.

Pond, Jean. "Mutual Humility in the Relationship Between Science and Christian Theology." In *Science and Christianity: Four Views*, edited by Richard Carlson, 67–104. Downers Grove, IL: InterVarsity, 2000.

Quan, Jacklin. "Scientists May Have Finally Figured Out How Elephants Got Their Incredible Trunks." *LiveScience*, December 6, 2003. https://www.livescience.com/animals/elephants/scientists-may-have-finally-figured-out-how-elephants-got-their-incredible-trunks.

Ratzinger, Joseph (Pope Benedict XVI). *Christianity and the Crisis of Culture*. San Francisco: Ignatius, 2006.

Rees, Martin. *Just Six Numbers: The Deep Forces That Shape the Universe*. New York: Basic, 2001.

Rose, Steven, et al. *Not in Our Genes*. London: Penguin, 1990.

Rose, Hilary, and Steven Rose, eds. *Alas, Poor Darwin: Arguments Against Evolutionary Psychology*. New York: Crown, 2000.

Rosenburg, Alex. *The Atheist's Guide to Reality*. New York: Norton, 2011.
Ruelle, David. *Chance and Chaos*. Princeton: Princeton University Press, 1991.
Ruse, Michael. "How Not to Solve the Science/Religion Conflict." *Philosophical Quarterly* 62.248 (2012) 620–25.
———. *Taking Darwin Seriously*. Amherst, NY: Prometheus, 1998.
Russell, Bertrand. "Reply to Criticisms." In *The Philosophy of Bertrand Russell*, edited by P. A. Schilpp, 643–78. Chicago: Northwestern University Press, 1944.
Russell, Robert. "Special Providence and Genetic Mutation: A New Defense of Theistic Evolution." In *Evolutionary and Molecular Biology*, edited by Robert Russell et al., 191–223. Rome: Vatican Observatory, 1998.
———. "What We've Learned from Quantum Mechanics About Non-Interventionist Divine Action in Nature—and Its Remaining Challenges." In *God's Providence and Randomness in Nature*, edited by Robert Russell and Joshua M. Moritz, 133–71. West Conshohocken, PA: Templeton, 2018.
Sagan, Carl. *Cosmos*. New York: Random House, 2002.
Schőnborn, Christoph. *Chance or Purpose? Creation, Evolution, and Rational Faith*. San Francisco: Ignatius, 2007.
Sellars, Roy Wood. *The Philosophy of Physical Realism*. New York: Russell, 1966.
Shanahan, Timothy. "The Evolution Indeterminism Thesis." *BioScience* 53 (2003) 163–69.
Silva, Ignacio. "Great Minds Think (Almost) Alike: Thomas Aquinas and Alvin Plantinga on Divine Action in Nature." *Philosophia Reformata* 79 (2014) 8–20.
———. "Thomas Aquinas on Natural Contingency and Providence." In *Abraham's Dice: Chance and Providence in the Monotheistic Traditions*, edited by Karl Giberson, 158–74. New York: Oxford University Press, 2016.
———. "Thomas Aquinas and Some Neo-Thomists on the Possibility of Miracles and the Laws of Nature." *Religions* 15.422 (2024) 1–14.
Sperry, R. W. *Science and Moral Priority*. Oxford: Blackwell, 1983.
Stump, Eleonore. *Aquinas*. London: Routledge, 2005.
———. *Wandering in Darkness*. New York: Oxford University Press, 2012.
Suppes, Patrick. "The Transcendental Character of Determinism." *Midwest Studies in Philosophy* 18 (1993) 242–57.
Sweetman, Brendan. *Evolution, Chance, and God*. London: Bloomsbury, 2015.
———. *Religion and Science: An Introduction*. London: Continuum, 2010.
———. *Why Politics Needs Religion: The Place of Religious Arguments in the Public Square*. Downers Grove, IL: InterVarsity, 2006.
Swinburne, Richard. *Are We Bodies or Souls*? Oxford: Oxford University Press, 2019.
———. *The Coherence of Theism*. Oxford: Clarendon, 1977.
———. *The Existence of God*. Rev. ed. Oxford: Oxford University, 1991.
———. *The Evolution of the Soul*. Rev. ed. Oxford: Oxford University Press, 1997.
———. *The Resurrection of God Incarnate*. Oxford: Oxford University Press, 2003.
———. *Revelation: Metaphor and Analogy*. Oxford: Oxford University Press, 1992.
Taylor, Charles. *Sources of the Self*. Cambridge: Harvard University Press, 1992.
Tracy, Thomas. "Particular Providence and the God of the Gaps." In *Chaos and Complexity*, edited by Robert Russell et al., 289–324. Rome: Vatican Observatory, 2000.
Turner, Derek, and Joyce Havstad. "The Philosophy of Macroevolution." In *Stanford Encyclopedia of Philosophy*, edited by Edward N. Zalta and Uri Nodelman. https://plato.stanford.edu/entries/macroevolution/.

Van Inwagen, Peter. *God, Knowledge, and Mystery*. Ithaca, NY: Cornell University Press, 1995.

Van Till, Howard. *The Fourth Day*. Grand Rapids: Eerdmans, 1986.

———. "The Fully Gifted Creation." In *Three Views on Creation and Evolution*, 159–247. Counterpoints. Grand Rapids: Zondervan, 1999.

Weinberg, Steven. *The First Three Minutes: A Modern View of the Origin of the Universe*. New York: Basic, 1993.

Wildman, Wesley. "The Divine Action Project, 1998–2003." *Theology and Science* 2 (2004) 31–75.

Wildman, Wesley, and Robert Russell. "Chaos: A Mathematical Introduction with Philosophical Reflections." In *Chaos and Complexity: Scientific Perspectives on Divine Action*, edited by Robert Russell et al., 49–90. Rome: Vatican Observatory, 2000.

Wiles, Maurice. *God's Action in the World*. 1986. Reprint, Eugene, OR: Wipf and Stock, 2010.

Willard, Dallas. *The Divine Conspiracy*. San Francisco: Harper, 1998.

Williams, Rowan. *Tokens of Trust: An Introduction to Christian Belief*. Louisville: Westminster John Knox, 2010.

Wilson Edward. *Consilience*. New York: Vintage, 2014.

———. *Sociobiology: The New Synthesis*. Cambridge: Harvard University Press, 1975.

Wright, N. T. *The Resurrection of the Son of God*. London: SPCK, 2003.

———. *Simply Christian*. San Francisco: HarperOne, 2021.

Index

www.ingramcontent.com/pod-product-compliance
Lightning Source LLC
LaVergne TN
LVHW100525110826
845146LV00002B/773
9798385261864